Disaster Emergency Management

Disaster Emergency Management

The Emergence of Professional Help Services for Victims of Natural Disasters

Liza Ireni Saban

State University of New York Press

Published by State University of New York Press, Albany

© 2014 State University of New York

All rights reserved

Printed in the United States of America

No part of this book may be used or reproduced in any manner whatsoever without written permission. No part of this book may be stored in a retrieval system or transmitted in any form or by any means including electronic, electrostatic, magnetic tape, mechanical, photocopying, recording, or otherwise without the prior permission in writing of the publisher.

For information, contact State University of New York Press, Albany, NY
www.sunypress.edu

Production by Ryan Morris
Marketing by Michael Campochiaro

Library of Congress Cataloging-in-Publication Data

Saban Ireni, Liza.
  Disaster emergency management : the emergence of professional help services for victims of natural disasters / Liza Ireni Saban.
     pages cm
  Includes bibliographical references and index.
  ISBN 978-1-4384-5242-5 (paperback : alk. paper)
  ISBN 978-1-4384-5243-2 (hardcover : alk. paper) 1. Emergency management. 2. Crisis management. 3. Disaster relief. 4. Human beings—Effect of climate on. I. Title.
  HV551.2.S23 2014
  363.34'8--dc23
                         2013030308

10 9 8 7 6 5 4 3 2 1

Contents

List of Illustrations — vi
Acknowledgments — vii
Introduction — 1

1. Introduction to Disaster Management — 13
2. Defining Disaster Vulnerability — 31
3. Vulnerability Assessment of Disaster Management Doctrines — 47
4. Applying Communitarian Social Justice in Public Administration Ethics — 61
5. Walzer's Communitarianism in the Service of Disaster Resilience Doctrine — 81
6. Comparative Analysis of Community-Based Disaster Resilience Policies — 103
7. Administration and Community Collaboration in Disaster Management — 165

Notes — 179
References — 195
Index — 225

Illustrations

Figure 1.1: The interaction between risk, susceptibility, capacity, and disaster event — 17
Figure 1.2: The disaster management cycle — 23
Table 3.1: Comparison of Disaster Management Doctrines — 58
Figure 5.1: The community-based disaster resilience model — 82

Acknowledgments

I wish to thank some of the people who contributed to this book.

I would like to thank Professor Alex Mintz, Dean of the Lauder School of Government, Diplomacy and Strategy at IDC Herzliya, who has been a constant source of support and inspiration. I wish to thank Professor Dave Nachmias, head of the Public Policy and Administration program at the Lauder School of Government, Diplomacy and Strategy, for his encouragement and continuous support through this project.

Many thanks to Dr. Michael Rinella, senior acquisition editor of SUNY press, who provided professional and enthusiastic support throughout this project. I would like to thank anonymous reviewers of the manuscript. They provided extremely constructive criticism that helped me to improve the quality of the book. I would also wish to thank Ryan Morris, senior production editor, and the production team at SUNY for their editorial assistance.

I am extremely grateful to Edna Oxman for her crucial editing and formatting assistance throughout the entire project. Edna has been a great source of encouragement and support.

I would like to thank my parents, Varda and Alfred. Without their unwavering support over many years, this book would not exist.

Last but not least, I would like to thank my husband, Jacob. Jacob has been a source of continuing love, patience, and support throughout the entire project. I also wish to thank my children, Amit and Lihi, who have been a source of motivation.

This book is dedicated to the memory of my grandmother, Alisa Mansharov. Not one single day goes by without missing her.

Introduction

The question of survival in times of disaster and trauma has always been of interest to laypeople, as well as to professionals and people in positions of authority. Indeed, throughout history we have seen how some societies have flourished in times of strife while others have not. We have seen this in times of natural disaster and in times of human-made disasters. What makes nations, communities, and individuals survive is one fundamental existential question with many different answers.

Strengthening community resilience in the face of or in preparation for adversity is one possible answer. According to the "Whole Community Approach," this has been a strategic goal entrenched in the U.S. Presidential Policy Directive 8 (PPD-8). The power of community resilience was well demonstrated in Japan in 2011, when a magnitude 8.9 earthquake struck off the coast, killing 15,790 people. According to the Japanese National Policy Agency, as of April 2011, the Japanese people responded to this terrible crisis with spirit and hope and stood strong despite such tragedy. Foreign journalists and scholars such as Gregory Pflugfelder, director of the Donald Keene Center of Japanese Culture at Columbia University, emphasized the role of local communities and civil organizations in manifesting effective disaster response efforts.

This book suggests that as long as this "bottom-upness" of disaster emergency management processes continues to persist, the capacity of policy makers and public administrators to approach emergency management with the kind of resilience needed is greatly undermined.

The originality of this book rests in bringing a new impetus to the role of public administrators as "professional helpers" in serving and empowering vulnerable communities (civic society groups) and thereby encouraging community resilience.[1]

This implies the need to focus on the normative and practical aspects of the forms of interactive governance networks between public administrators and affected communities in disaster response and recovery efforts. This book will highlight how public administrators, namely local and street-level public officials, should strengthen their communities' capacity to meet their shared interests as a proactive measure ensuring very basic community resilience capacity building, including strengthening a sense of belonging, trust, and collaboration with and among members of affected communities.

Disaster response and recovery are often characterized by nonhierarchical networks that cross many sectors. Research on governance networks in emergency management was developed in critical opposition to the traditional focus on government and top-down steering. Many of the most innovative and effective initiatives to address disaster relief and recovery efforts have been designed and implemented by civil society groups—international nongovernment organizations (INGOs), nongovernment organizations (NGOs), and local groups.[2] Thus, the reason for devoting much attention to the "bottom" was not to side with underdogs, but to arrive at similar understandings of disaster management from a different perspective, one that is especially conductive to exploring the interactive networks that function in response and recovery processes, in which various actors and organizations take part. However, governments and public administration are reluctant to grant too much credibility to civil society initiatives. It is often claimed that leaving disaster management in the hands of civil society organizations and local groups may open the possibility that those groups will broaden their agendas and question the priorities and methods of the elite-controlled political processes.[3] Moreover, it questions the value of disaster management agencies to design, administer, and evaluate policy responses if better work can be done by civil society organizations. Civil society organizations may not have complete understanding of the wider socioeconomic context in which proposed actions would occur.

For that, the present book argues for asymmetric interdependencies, which means that some actors are more central than other actors. Devoting attention to the "top" is not to privilege elite policy makers but to make normative and practical diagnoses of how disaster management can be improved. Thus, this book offers to take a bottom-up view of action in governance networks even where dominant disaster decision making taken at the "top" sets the framework for action lower down. The state is indeed "brought out" in this mixed model, yet theorizing that the state should be concerned with the relationship between the state and civil society in disaster management.[4]

The emphatic plea that the importance of the state (hierarchical form that underlies top down) must be incorporated into governance networks is made by the juxtaposition of network and hierarchic forms of governance. The possibility of mutually reinforcing interaction between the state and civil society to create more power on both sides in specific situations depends on the feasibility of institutions that can link state power with social forces. Thus, the present book suggests that public administration can facilitate the possibility of mutual empowerment through collaboration with active civil associations in times of natural disaster.

When natural disasters occur, public officials, namely local and street-level administrators rather than the government itself, find themselves on the front line in dealing with complex public problems. As a result, public officials act as the middle man interacting with government officials, communities, nongovernmental organizations, and the private sector. This is not an easy task in "good times" and certainly not in times of crisis and disaster. Recent studies on disaster management, rescue, and relief measures have reinvigorated heated debates and concerns over the breakdown of government and bureaucratic decision-making in the face of pressing need.[5]

These studies show how contested ethical issues arise when public administrators use a technical-rational approach when forced to comply with authorities during policy implementation. This book will redefine the role of public administrator to alleviate disaster vulnerability and thereby encourage community resilience using a partiality approach.

Based on such deliberations, public servants need to assure that response and recovery efforts are fully consistent with norms of justice and fairness and take an active role in creating opportunities in which affected communities can articulate shared values and collective interests rather than attempt to comply with strict rules and procedures of distribution. Thus, the focus of the normative discussion is on both skill development and personal characteristics that enable civil servants to serve the public by helping vulnerable communities meet their shared interests and coping capacities in an effective and equitable way, rather than steering. There is no one set of characteristics that identify "professional helpers," but this book encourages public officials to think about the characteristics they possess that could either help or hinder them in their work with others.

What does this book contribute to the field of Disaster Management?

This book suggests that reexamining and modifying the role of the public administrator in disaster management holds an important disciplinary and practical justification. First, it advances government and administration understanding and handling of these calamitous events, which can lead to improvement in the lives and livelihood of citizens in society. Beyond this practical reason, however, there is an important disciplinary reason for studying the role of public administration in the face of adversity. Disasters challenge the very nature of public administration. Disasters highlight one of the major distinctions between instinctive and deliberate (bureaucratic) action. Thus, disasters provide unique opportunities for learning more about the routine and the normal functioning of societies. Abnormal phenomena disrupt the accustomed routines of daily functioning, especially the characteristics of routine and predictability of working in a bureaucracy. This book will show how the threat of disasters should and could be used as opportunities for taking an active role and developing professional skills to build relationships of trust and collaboration with and among vulnerable citizens that link the professional with the private person of the civil servant.

Most of the literature in this domain addresses distribution services for basic hard provisions, such as accommodation, sanitation, food, and water sources. This book suggests that these potential negative consequences go beyond loss of life and damage to property and bring into the fore notions of fairness and social justice. It is argued that the way citizens are treated in the delivery of goods by public administration (e.g., local and lower-level administrations) in times of emergency has a great impact on citizens' trust in governance arrangements. Local officials must be sensitive and modify their approach in order to identify affected communities' vulnerabilities and concerns. If not done, citizen evaluations may not match objective output measures of services precisely because they do not include the interpersonal aspect of service delivery so critical to the end user.

This book shows how the relationships between public officials and communities in times of natural disasters evolve and are shaped in order to reduce vulnerabilities. Four different settings, the Gulf Coast Hurricanes (United States), the West Sumatra Earthquakes (Indonesia), the Wenchuan Earthquake (China), and the Great East Japan Earthquake, will be explored. The comparison between these cases provides insights into the distinctive role of public administration as professional helpers in disaster emergency management.

In order to evaluate the normative role of administrative agencies in different disaster management settings, we use a data set collected from archival documents, describing organizational collaborations between public and civil organizations engaged in resilience efforts. The collected data for this study issued from content analysis of news reports, governmental bills, proposals, statements, press releases, position papers on the subject, testimonies at government hearings, committees' consultation responses, and situation reports (accessible through organizations' websites during the course of a disaster) using Lexis-Nexis program. Archival data is preferred for the present study since it can be available for an extended period of time and in situations in which data collection is difficult. The data utilized in this book provides lists of organizations that responded to the event, and instances and patterns of collaborations among them.

How is this book theoretically and practically attractive to the study of disaster management?

This book is born from my curiosity about the normative evaluation of the disaster management outputs. The intellectual literature of the study of public administration has tended to cloud the subjective component of disaster management. The way citizens are treated in the delivery of goods by public administration, and especially the local and lower levels of that administration that interact with citizens and groups, has a great impact on the role of and the respect for political systems in the society.

The argument is that in times of emergency, citizens may not have the luxury of finding alternative ways to supply needed goods or engage in assessment of the fairness and effectiveness of service delivery. Viewed in this way, administration does make policy, as administrators hold accountability to assure that provided services and goods are consistent with norms of justice and fairness. Thus, the way affected communities are treated in the delivery of goods by public officials (e.g., local and lower-level administrations) in times of emergency has a great impact on citizens' trust and participation in governance arrangements.

The idea behind this book is to illuminate the ways public administration may address community and civil society in an environment of disasters. Among others, it raises the question of whether public administration should work to encourage citizen identification with community or the civil society at large. I want to turn that focus, and the related question, that civil society, or more specifically community, can provide public administrators a means of dealing with the wide range of challenges posed by the disaster events. In times of adversity, the sense of relatedness, shared responsibility, and solidarity is increasingly downplayed, but its function is critical for any sort of effective collective action to take place. Thus, the subjective component of the strength of performance of public administration will be expected to contribute to building civil society capacities by advancing citizen identification with community.

For disaster management to deal fairly with the causes of disadvantage faced by disaster-affected communities, communities and the general society need to accept disaster policy interventions as effective and just. This implies quite a change in the ordinary mode of disaster administration-less attention to what to do when disaster strikes, and more attention to the issue of how government and administrations strengthen their communities, which means very basic capacity building, including strengthening of the social bond. The concept of vulnerability has become an important part of natural disaster and food security analyses since the 1980s. Focusing on vulnerability has been seen as a way to improve disaster mitigation by shifting emphasis from natural causes to the social processes that cause some people to be more greatly affected than others. Recent models of disaster risk management seek to adopt a vulnerability approach that is concerned with reducing the vulnerability of communities. Within the context of vulnerability-based disaster risk management policies, much research has concentrated on the causes of vulnerability to disaster. However, there has been relatively little research that sees disaster vulnerability issues in the process of implementing emergency policies by public administration as worthy of investigation. For this purpose, I apply the resilience approach to profiles of community-administration interactions for better implementation of disaster management policies that underlie the dynamic community-environment interactions affecting the adaptive responses to emergency.

Consistent with the understanding of the role public administration plays in times of disaster as improving interpersonal and social processes as "professional helpers," lies the salient features of participation that invite various stakeholders from affected communities to share the responsibility of disaster emergency policy implementation. Needless to say, inclusive and participatory framework is not a panacea for all risk management problems. A competent, accountable, and equitable agency deliberation is still better than a superficial consensus among affected parties. The process of becoming a helper is intrinsically related to personal development and the ability to empower others. The focus of this discussion is on both skill development and personal characteristics that enable civil servants to be effective and just. There is no one set of characteristics that identifies "professional helpers," but we encourage public officials to think about the characteristics they possess that could either help or hinder them in their work with others. This book intends to provide public officials with some characteristics that are worthy of reflection such as advocacy, inclusion, and competency. Toward this end, we explore how these values operate in helping relationships.

Given the subjective and normative assumptions underlying a resilience approach to community-based disaster management as to what constitutes positive or desirable outcomes for resilient communities, the question at issue here remains how to validate the application of social justice of specific administrative choices and practices, along with their efficiency consequences. This book urges that studies of disaster management be explicitly and systematically grounded in core theoretical concerns of applied ethics, and argues that Michael Walzer's communitarian view of social justice provides one way to do this.

Walzer's theory of social justice is particularly relevant in light of the central role community occupies in disaster programs and policies. In this sense, the Walzerian theory of social justice is morally instrumental in a relationship to the extent that it contributes to the protection of those who are in need and on social conventions that assign responsibilities for the care of needy persons to others who stand in certain relationships to them.

Communitarian justice criteria have the potential to add a new dimension to the understanding of public administration practice in the recent Gulf Coast Hurricanes (United States), the West Sumatra Earthquakes (Indonesia), the Wenchuan Earthquake (China), and the Great East Japan Earthquake. On December 26, 2004, Indonesia experienced the 2004 Sumatran earthquake. About 170,000 people were reported dead, with an estimate of more than 37,000 missing. In addition, it was reported that hundreds of buildings had collapsed, which left thousands of people homeless. On August 29, 2005, the center of Hurricane Katrina passed east of New Orleans; winds downtown were in the Category 3 range with frequent intense gusts and tidal surges. At least 1,836 people lost their lives and 80 percent of New Orleans was flooded, with some parts under 15 feet (4.5 m) of water. Another deadly disaster occurred on May 12, 2008, in the Sichuan province of China (also known as the Wenchuan Earthquake). At least 69,000 people were killed, 374,176 injured, with 18,222 listed as missing; about 4.8 million people were left homeless, as reported by the Ministry of Civil Affairs of the People's Republic of China. A more recent case that is dealt with in this book is the Tōhoku earthquake and tsunami that occurred in Japan on March 11, 2011. Japan, in effect, suffered three disasters on March 11, 2011. There was an offshore earthquake that, given its scale, caused relatively little damage on the mainland and only about 10 percent or less of the total number of deaths in the overall event. There was a tsunami that destroyed part of the settlements and killed about 90 percent of the total number of deaths. The Japanese National Police Agency has reported 15,760 deaths, 5,927 injured, and 4,282 people missing. In addition, over 125,000 buildings were damaged or destroyed. The disaster also led to nuclear accidents at three reactors in the Fukushima I Nuclear Power Plant complex; leaks led to a 30 kilometer evacuation zone surrounding the plant. By studying these examples, it is possible to examine the extent to which governmental and administrative deliberations in disaster relief efforts in these regions meet resilience principles, their degree of success, and the limits of these efforts.

Since this book addresses a community-based disaster risk approach to society-administrative relations, the term *civil society* is used to distinguish between autocratic and democratic regimes, and refers to the community of citizens within a country that is actively engaged in governing the country within legally defined limits. The U.S. context represents an old and mature civil society; the Indonesian case represents a young civil society, which until 1998 experienced dependence on an authoritarian regime; the Japanese case represents a semiindependent civil society that is often allied to or collaborates with the state; and the Chinese case represents a greatly dependent civil society in an autocratic state. Thus, the comparison of these cases meets an undeniable interest that exists in the role of public administration in disaster emergency management as part of the process of building civil society capacity in Asia and thereby plays a useful role integrating our world just a little bit more, as disasters clearly affect the lives and livelihoods of humans all over the world. In the U.S. setting, the interactions between affected communities and public officials were quite polarized; in Indonesia and China, communities' members and public officials were more committed to contribute to building collective shared perceptions of need and relationships of collaborations. Higher levels of local competency and enthusiasm facilitated disadvantaged communities' inclusion and motivated the cooperation between communities and external organizations, such as voluntary and local business organizations to promote the effectiveness of community adaptation in the face of disaster risks in Indonesia, and to a greater extent in China. The Chinese case as well as the recent disaster event in Japan highlight the importance of less diversified society where both public officials and communities are familiar with daily practical needs, physical needs, and maintenance of social norms and order in developing spontaneous interactions and collaborative efforts, with public administration tasked to respond in an effective and equitable manner to the disaster risks. Finally, this book aims to reach a diverse and broad audience of practitioners, policy makers, civil servants, disaster managers, scholars, students, or simply those with curiosity about natural disasters as a social laboratory and catalyst to strengthen civil society.

Chapter-by-Chapter Synopsis

Chapter 1: Introduction to Disaster Management

Learning Objectives

On completing this chapter you will be able to

- describe in general, ancient, and modern disaster management programs;
- define the theoretical underpinning and foundation of disaster management;
- discuss basic approaches to disaster management (i.e., top-down and bottom-up);
- identify the complexity of the disaster management process by breaking it into stages of mitigation, preparedness, response, and recovery;
- discuss how policy formulation in disaster management is determined; and
- list examples from communities showing how policy formulation for disaster management was accomplished.

Chapter 2: Defining Disaster Vulnerability

Learning Objectives

On completing this chapter you will be able to

- define various areas of vulnerability (e.g., food, accommodation, and human security analyses);
- describe what makes a community vulnerable in times of disaster;
- discuss if there is there a unified conceptualization of community vulnerability;
- describe approaches for addressing vulnerability in different settings;
- list and evaluate three main approaches to vulnerability and risk reduction in disasters or hazardous settings, such as resistance, sustainability, and resilience, referring to the means being employed and the objectives being served; and
- delineate how vulnerability should be transformed into a more meaningful and usable construct for our understanding of the social impact of natural disasters.

Chapter 3: Vulnerability Assessment of Disaster Management Doctrines

Learning Objectives

On completing this chapter you will be able to

- establish a policy evaluation framework of the relations between instruments and their effects in the field of disaster management;
- discuss the effectiveness of existing policy instruments for disaster management;
- assess the extent to which disaster management policy instruments have achieved their stated goals in international disaster management settings; and
- encourage a critical reflection of policy instruments' effects and stated goals to identify areas of strength and weakness and ways to improve the performance of disaster management practices.

Chapter 4: Applying Communitarian Social Justice in Public Administration Ethics

Learning Objectives

On completing this chapter you will be able to

- describe an improved model for disaster management that includes a set of variables that affect disaster management;
- discuss how two distinct but complementary theories can operate to establish and build the social bond between public administration, civil service, and society; anddiscuss how Walzer's communitarian justice principles impact community and public administration relations.

Chapter 5: Walzer's Communitarism in the Service of Disaster Resilience Doctrine

Learning Objectives

On completing this chapter you will be able to

- discuss the principles underpinning a resilient community by applying Walzer's theory of social justice; and
- delineate a comprehensive community perspective on disaster management to identify the way public administration and

community should interact, involving three levels of intervention: advocacy, inclusion, and competence.

Chapter 6: Comparative Analysis of Community-Based Disaster Resilience Policies

Learning Objectives

On completing this chapter you will be able to

- describe the implication of disaster management policies by using as examples the Gulf Coast Hurricanes in the United States (2005), the West Sumatra Earthquakes in Indonesia (2005), the Great East Japan Earthquake (2011), and the Wenchuan Earthquake in China (2008).
- discuss how advocacy, inclusion, and competence are attributes that emphasize deep social and community transformation rather than merely helping people adapt to risks and crisis circumstances.

Chapter 7: Administration and Community Collaboration in Disaster Management

Learning Objectives

On completing this chapter you will be able to

- describe the new "professional helper" as a role in the face of adversity;
- delineate what set of characteristics they need both personally and professionally to be the best helper possible;
- discuss how to encourage public administrators to reflect on the characteristics they possess that can either help or hinder them in their resiliency efforts and work with affected communities;
- delineate what it takes to implement this role;
- list the tasks of the professional helper; and
- design a scheme for community-based disaster management in which both administration and communities must recognize and perform their roles in large cooperative efforts toward the development of resilient communities.

1
Introduction to Disaster Management

A question that may be raised is why study disaster at all? A good answer may be found in the novella written by Heinrich von Kleist, *The Earthquake in Chile* (1807). The novella draws on two disaster events: the 1647 earthquake in Santiago, Chile, and the in 1755 earthquake in Lisbon, Portugal. The author used these events as a social laboratory to examine whether they lent little if any support to the hypothetical State of Nature, which was praised by the eighteenth-century political philosopher Jean-Jacques Rousseau as a normative guide of "uncorrupted morals." The State of Nature was used as a thought experiment to develop the hypothetical conditions that preceded authority by consent or governance. The disaster was supposed to serve as a "state of nature" to build a society with no institutionalized religion or government, which led to the unavoidable conflict between individual morality versus society's conventions. As Kleist described it,

> In the minds of Jeronimo and Josefa strange thoughts began to stir. When they found themselves treated with so much familiarity and kindness they did not know what to think of the recent past: of the place of execution, the prison and the bells; or had all these been merely a dream? It seemed that in everyone's mind, after the terrible blow that had so shaken them all, there was a spirit of reconciliation. Their memories seemed not to reach back beyond the disaster.

However, during the recovery efforts, survivors soon became nasty, brutish, and egoist. No cooperation or empathy came from the momentum of the disaster.

An answer to the question previously raised seems obvious in this context. Disasters can be found to produce cultural, social, economic, and psychological consequences for individuals and communities.[1] Viewed in this way, disasters provide opportunities to reflect on social structures and processes that lie behind daily functioning of societies, since disasters are "nonroutine events in societies . . . that involve conjunctions of historical conditions and social definitions of physical harm and social disruption."[2] Thus, an abnormal phenomenon offers the means for identifying certain normal features of the structure and functioning of societies. In short, the study of disaster as disruption of routines and threats of disruption is a well-established tool for advancing the understanding of the mechanisms that build and rebuild personality and social structures.[3]

Thus, the second practical justification for the study of disasters lies in our very nature to enlarge our understanding of these catastrophic events in order to be able to lessen their devastating consequences. Disasters then function as a catalyst for collective action that permeates a community's social structure, producing social responses that are both emergent and constraining.

Definitions of Disaster

In recent decades, it has become increasingly clear that large-scale disasters will be persistent features of social life. According to the 2010 World Disaster Report, natural disasters of the last fifty years are taking tolls in human life, property damage, and social and economic disruption.[4]

The root of the word *disaster* is derived from Greek astrological study in which this term was used to refer to a destruction or deconstruction of a star "dus-aster" ("bad star"). Disaster is defined as a sudden event causing great damage and loss of life and property that far exceeds our capabilities to recover.

Although there is little consensus among scholars on the definitions of *disaster*,[5] Quarantelli offered a comprehensive definition that bears on the shared defining component of disaster, that is, of the negative consequences of the disruption of the accustomed routines of daily functioning at the collective level. According to Quarantelli, disasters are

> those crisis occasions generated by the threat of or the actual impact
> of relatively sudden natural and technological agents (such as

earthquakes, floods, hurricanes, volcanic eruptions, tornadoes, and tsunamis as well as toxic chemical spills, radiations fallouts, large-scale explosions and fires, structural failures, massive transportation wrecks and crashes, etc.) that have significant negative social consequences. Basically we include only those instances where everyday community life is disrupted and where local resources cannot handle the demands of the situation.[6]

The United Nations International Strategy for Disaster Reduction (UN/ISDR) suggests viewing disaster:

> A disaster is a sudden, calamitous event that causes serious disruption of the functioning of a community or a society causing widespread human, material, economic and/or environmental losses which exceed the ability of the affected community or society to cope using its own level of resources.[7]

Such definition focuses on the immeasurable losses caused by disasters, which vary with regional location, climate, and the degree of vulnerability. For a disaster to be considered under the database of the UN's International Strategy for Disaster Reduction (ISDR), at least one of the following features must be met:

- a report of 10 or more people killed;
- a report of 100 people affected;
- a declaration of a state of emergency by the relevant government; and
- a request by the national government for international assistance.

Most definitions of disaster include common main features, such as unpredictability, unfamiliarity, promptness, urgency, uncertainty, and hazard. Based on these shared characteristics, we can define disaster as a hazard causing great losses to life, property, and livelihood, and uncertainty.

Types of Disasters

Disasters can be distinguished as natural disasters, often regarded as "acts of God," such as earthquakes, floods, landslides; and man-made disasters or technological disasters, such as war, bomb blasts, and chemical leaks. However, such typology has resulted in continuous debates among scholars. Those who claim a distinctive nature of human/technological disasters bring the evidence that "Technological disasters create a far more severe and long lasting pattern of social, economic, cultural and psychological impacts than do natural [disasters]."[8] Others argue that such a distinction is theoretically and practically specious, since disasters have no single root cause and result from human failure to introduce appropriate emergency-management measures.[9]

Although this debate still continues, it could be agreed by all that disaster caused by nature can have human origins. Natural disasters can result from the combination of a hazardous environmental process, susceptibility of a given population to that process, and inability to mitigate the potential negative consequences of this process when assessed in human terms. It follows that natural disasters often differ in quantity of damage caused or in quality of the type of negative consequences. Moreover, natural disasters that impose a great risk to one particular society may not be assessed in a similar way by a different society with different susceptibility and capacity features (see fig. 1.1).

Natural Disasters

Based on the outlines of the UN Office for the Coordination of Humanitarian Affairs, natural disasters can be divided into three main groups: hydrometeorological, geophysical, and biological.

- Hydrometeorological disasters originate from natural processes or phenomena of an atmospheric, hydrological, or oceanographic nature that may lead to personal injuries or losses of life, property damage, social and economic disruption, or environmental degradation. These include floods and wave surges, storms, landslides, avalanches, and droughts and related disasters (extreme temperatures and forest/scrub fires).
- Geophysical disasters are basically earth processes or phenomena that may also cause loss of life or injury, property damage, social and economic disruption, or environmental degradation. These include earthquakes, tsunamis, and volcanic eruptions.

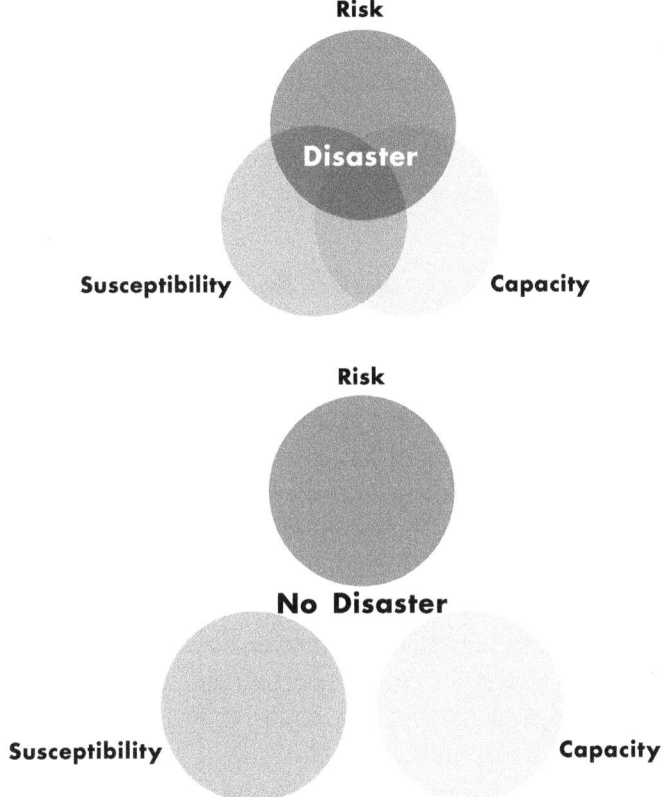

Figure 1.1: The interaction between
risk, susceptibility, capacity, and disaster event

- Biological disasters are those processes that originate from biological vectors, including exposure to pathogenic microorganisms, toxins, and bioactive substances, which may cause loss of life or injury, property damage, social and economic disruption, or environmental degradation. These include epidemics and insect infestations.

Common to all types of natural disaster is the fact that the social and economic disruptions usually impose direct (e.g., damage to infrastructure, crops, housing) and indirect (e.g., loss of revenues, unemployment, market destabilization) impact on the local economy and social structure. However, disasters are not totally discrete phenomena. Their occurrence, time, place, and intensity could be predicated to some extent in some cases by technological and scientific means. Thus, we can expect to retain at least some of the capacity to reduce the impact of disasters by adopting suitable mitigation means, though we cannot reduce the extent of damage and loss itself.

The History of Disaster Management

Ancient disaster management programs

Disaster emergency management is a modern discipline of dealing with and avoiding risks imposed by natural catastrophes such as fire, flooding, or earthquakes. However, ancient societies showed signs of disaster emergency programs. Before the eruption of Mount Vesuvius in 79 AD, local inhabitants from the districts of Rome, as well as historians who lived through the period, did not report earthquakes and fires since these disruptions were regarded as common features of social life at that time. Even when eruptions were reported, such as the Great Fire of Rome that began July 64 AD and was reported by Tacitus,[10] they were often conceived as rumors and unreliable.

Local historians reported that the fire of 64 AD caused the destruction of three districts of Rome and ten other cities suffered serious damage.[11] Emergency relief efforts were organized and funded by Nero, the emperor at the time.[12] During the fire, Nero opened his palaces to provide shelter for the inhabitants who lost their houses during the fire, and organized a chain of food delivery to prevent starvation among the survivors.[13] After the fire, Nero planned a new urban development program that included new building rules, such as spacing houses and to face porticos on wide roads. It is interesting to note that Nero's admirable relief efforts in the aftermath of the fire were debated by the historians of the time (and by Modern historians as well), as some say that it was Nero himself who set the fire to build himself a new palace complex, which is why he rushed to execute relief programs.

The eruption of 79 AD destroyed Pompeii, Herculaneum, Oplonti, and Stabiae. The Mount Vesuvius earthquake spread tons of molten ash. This eruption was recorded by Pliny the Younger, whose letters to his friend Tacitus provide an authentic description of the disaster:

Ashes were already falling, not as yet very thickly. I looked round: a dense black cloud was coming up behind us, spreading over the earth like a flood. . . . There were people, too, who added to the real perils by inventing fictitious dangers: some reported that part of Misenum had collapsed or another part was on fire, and though their tales were false they found others to believe them. A gleam of light returned, but we took this to be a warning of the approaching flames rather than daylight. . . . I could boast that not a groan or cry of fear escaped me in these perils, but I admit that I derived some poor consolation in my mortal lot from the belief that the whole world was dying with me and I with it.[14]

Relief efforts after the eruption were poor as most of the cities remained buried and undiscovered until excavation began during the eighteenth century. However, some recorded efforts were made by Emperor Titus, who appointed two expert counsels to manage the restoration plans of the damaged area.[15]

Selected Modern Disaster Management Programs of the Twentieth Century

The San Francisco earthquake of 1906

The 1906 San Francisco earthquake caused the largest urban fire in U.S. history. The earthquake and ensuing fire resulted in more than 3,000 deaths and the destruction of 492 city blocks.[16] The city's fire chief, Dennis T. Sullivan, who was injured and later died from his injuries, realized that his men were untrained in the use of dynamite to demolish buildings to create firebreaks. For that, together with San Francisco's Mayor Schmitz, he called army troops (over 4,000 men) to assist in the relief efforts and in using dynamite to demolish buildings. The army played a great role in disaster relief as it became responsible for supplying food, shelter, and clothing to tens of thousands of homeless residents of the city. The army established eleven temporary camps, including 5,610 redwood and fir "relief houses" to accommodate 20,000 displaced people. During and after the earthquake, the residents who survived were homeless, and were maintained in place by receiving specific instructions on digging latrines in backyards and providing water in tankers parked on street corners. However, this empirical and historical evidence should not be taken for granted. According to Quarantelli's critical discussion of statistical and empirical data on disasters,[17] the historians, Hansen and Condon[18] showed by careful analysis of the apparent prompt actions taken by the army and the local government leadership, that these efforts were not without criticism. In fact, allegations of political corruption and discriminatory practices to exclude Chinese residents were often said to play a role in recovery and reconstruction efforts.

The Yangtze River Flood in China of 1931

During late 1930, heavy snowstorms caused a series of floods during the Nanjing decade in the Republic of China era. The Yangtze River flood killed about 145,000 and affected 28.5 million residents.[19] (Several sources argue for 3–4 million deaths.) The relief efforts, mainly by local organizations, began shortly after the flooding became destructive. In Hankou, local residents raised 800,000 yuan to fund relief efforts and set up temporary relief camps that served 300,000 people.[20] The provincial reconstruction commissions as well as national and international organizations provided relief assistance. During the disaster, the National Flood Relief Commission (NFRC) was initiated to provide coordination and constructive solutions for the disaster's effects. The members of the commission were mostly governmental who deliberately designed a cooperation program to control international support from abroad.

The Great Alaskan Earthquake and Tsunami of 1964

The Great Alaskan Earthquake and Tsunami of 1964 caused the death of 131 people and immense destruction that was estimated at over $310 million.[21] After the earthquake, the West Coast and Alaska Tsunami Warning Center was established to monitor seismic activity and to broadcast to the public, triggering alerts to local, state, and federal emergency officials, including the military and the Coast Guard. Other reconstruction efforts were held by the State of Alaska, the U.S. Army Corps of Engineers, and the federal government as a whole to rebuild roads and completely destroyed villages such as the native village of Chenega and the town of Valdez.

The Bangladesh Cyclone of 1970

In 1970, the Bhola cyclone struck East Pakistan (now Bangladesh) and India's West Bengal. The tropical cyclone caused the death of 500,000 and great damage to villages and crops throughout the region. The Pakistani government was strongly criticized for its delayed handling of the relief efforts following the storm, both by local political leaders in East Pakistan and in the international media. Although the Indian government received many reports from ships containing meteorological information on the cyclone from the Bay of Bengal, such information was not passed on to the Pakistani government due to the rivalry in relations between India and Pakistan, costing thousands of lives.[22]

After the storm, the Pakistani army used gunboats and a hospital ship to carry medical personnel and supplies for the damaged islands of Hatia, Sandwip, and Kutubdia; only one military transport aircraft and three crop-dusting aircraft were assigned to relief work by the Pakistani government. However, the government neglected to coordinate with international and national organizations. For example, the Pakistan Red Crescent decided to operate independently of the government as the result of a dispute that arose after the Red Crescent took possession of twenty rafts donated by the British Red Cross.[23] Moreover, the Pakistani government did not allow the Indians to send supplies into East Pakistan by air, forcing them to be transported slowly by road instead. The Indian government also criticized the Pakistanis for refusing to deploy military aircraft, helicopters, and boats from West Bengal to assist in the relief operation. The hostile relations between the two countries that were intensified during the disaster management efforts helped to trigger the Indo-Pakistani War of 1971 in December and concluded with the creation of Bangladesh.

The Tangshan Earthquake in China of 1976

The Tangshan earthquake of 1976 in China caused the death of at least 255,000 people. Before the quake, the county of Qinglong was prepared for the quake two years earlier, as the county officials engaged in periodic emergency meetings to prepare and instruct villagers to evacuate to safer areas when the earthquake struck. Although preparatory measures were taken, great loss of life caused by the earthquake was attributed, among other causes, to the low quality and nature of building construction in China. During and after the quake, the Chinese government refused to allow foreign aid from the United Nations, the United States, or the Red Cross.[24] The Chinese government kept its self-reliance, and sent several medical teams to Tangshan in addition to the People's Liberation Army, who were assisting and engaged in rebuilding infrastructure immediately after the quake in Tangshan; the city was completely rebuilt.

Following the brief review of the selected emergency and relief efforts introduced in various natural disasters, it is argued that natural disasters can become the trigger of civil unrest and social and political criticism as the political, social, and technological environments are caught by the urgency and uncertainty of events or possible outcomes. Thus, key to disaster management is sensitivity to aspects of the dynamic between political and social institutions, in both managerial and normative terms. Although it is common to relate the main responsibility for emergency management to government agencies, emergency management is an integrative and complex process involving individuals, groups, communities, and professional scientific agencies. Thus, an effective emergency management results from the integration of emergency plans at all levels of government as well as nongovernment involvement (individual, group, and community).

Disaster Management Process

The disaster management process is defined as the possible actions taken by an organization to reduce the impact of disasters on humans, the built environment, or both. Although there is no agreed formula at the global level for how modern disaster management should be established and implemented, the following three aspects are mostly shared by distinctive disaster management programs:

- preparation for a disaster before it occurs by developing early warning devices;
- development of disaster response (e.g., emergency evacuation, and quarantine, mass decontamination); and
- support and rebuilding plans after natural or human-made disasters have occurred.

The Cycle of Disaster Management

The disaster management cycle can be broken into five main stages and phases of applied problem-solving as illustrated in fig. 1.2.

In this model, *disaster stage* refers to the process by which the event of the disaster takes place. The damage/loss of human life, loss of property, loss of environment, loss of health, and anything else is assessed by government agencies. This stage raises uncertainty and profound shock. The *response stage* refers to the process by which governmental and nongovernmental organizations, individuals, and communities respond to the disaster through first aid provisions such as food, medical aid, shelter, and counseling. This stage is likely to include first aid emergency services such as firefighters, police, and medical and ambulance crews. These organizations may be accompanied by secondary emergency services such as specialist rescue crews, NGOs, and local government agencies. The *recovery stage* involves rebuilding the affected area after immediate needs were met by the previous stage. This stage provides an opportunity for adopting social security efforts directed at restoration of damaged property, supplement of employment and educational solutions,

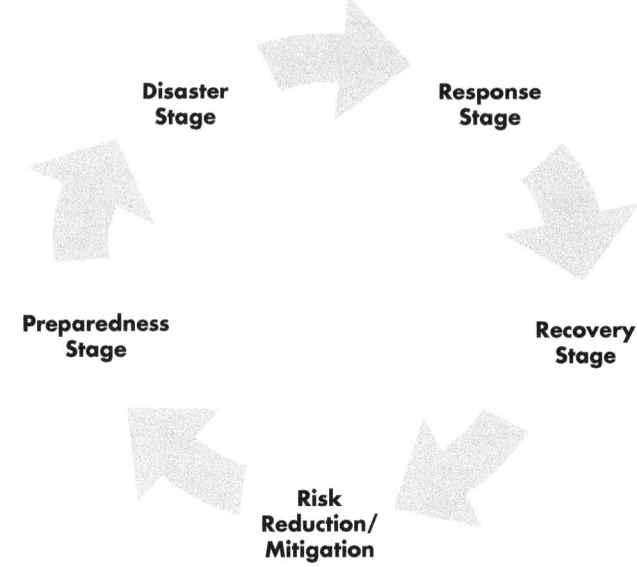

Figure 1.2: The disaster management cycle

and rebuilding essential infrastructures. The *risk reduction/mitigation stage* refers to the process by which the results of the recovery policies are monitored by both state and societal actors, the result of which may be reconceptualization of emergency management problems and solutions. For example, damages to property caused by an earthquake would lead to rebuilding resistant houses according to proper construction standards. In the case of tsunami, mitigation efforts would lead to preventing destruction of housing located close to the shore and the implementation of security measures such as a green belt—a thick, wide growth of trees bordering the coastline in order to reduce the impact of the tsunami waves on the land. During this stage, both governmental agencies and affected communities consider long-term measures for reducing the extent or impact of damage during the next similar disaster. This stage is thus built heavily on risk identification in order to address proper technological and planning devices based on probability and the level of impact of specific risk. In the *preparedness stage*, causality prediction and early warning options are formulated by the government. For this stage to be effective, the emergency management program of actions should cooperate with volunteers and effected communities to develop local capacities and coordination with government emergency teams when the disasters strike. The common measures used to mobilize resources are communication and telemedicine services, multiagency coordination, maintenance and transportation for emergency services, and local training of at-risk communities of warning measures, emergency shelters and evacuation options, and preparation of evacuation kits, commonly referred to as a 72-hour kit, which includes food, medicine, flashlights, candles, and money. Preparedness activities can be viewed as complementary to activities taken in the mitigation stage. For that, preparedness activities involve building coordination between the government and the private sector and nongovernmental organizations to improve prompt and efficient response and recovery efforts. It should be noted that in homeland security there is a clear preference to split the phase of risk reduction/mitigation to "prevention" and "protection" phases[25] in referring to these activities.

An important advantage of this disaster management model as set out above is that it facilitates understanding of the disaster management process by breaking the complexity and uncertainty created by disaster into a limited number of stages, each of which could be examined alone or in terms of its focus, measures, institutions, and actors. This model can be applied for comparative studies of disasters occurring in different regional settings or different stages of a given disaster; in other cases it can support the distinguished nature of the disaster decision-making process in comparison to the "normal" policy-making cycle. In addition, the model can denote the role of the various actors involved in disaster management, not just governmental agencies, which formally take over the emergency efforts. However, this model does have certain shortcomings. One major disadvantage is that while the logic and consistency of the model may be sufficient in theory, in the real world, especially in times of adversity, stages are more often skipped or followed in a different order than that specified by the problem-solving phases.[26] In addition, this model offers no indication of who or what prompts disaster emergency to progress from one stage to another, an issue of importance especially for scholars working on emergency planning programs. This model also suffers from lack of causation as to which factors underlie the process and may lead to certain emergency management decision making. This model contains a rather simple description of activities that occur in mitigation and response to disaster. In fact, Waugh[27] has offered to replace the use of "phases" with "activities" and "functions," which involve different types of professional expertise and skills as well as different types of agents. In this book we prefer to focus on agents involved in the emergency process. Policy making involves a multitude of actors, which can vary depending on how a vulnerability is defined, and facilitates the adoption of certain solutions to it. The next chapter seeks to capture the complexities by building deeper questions into the model and draws on the terms and concepts of contemporary political science in answering them.

Approaches to Disaster Management Policy Formulation

This section discusses the existing approaches to emergency management. As seen, studies on disaster management emphasizing policy formulation that has gone through a number of stages including preparedness, planning, mitigation, and recovery, reflect two well-known approaches–top-down and bottom-up.

The top-down approach refers to decisions made at the central-state level, and regulations are imposed in an exercise of top-down authority. This approach "assumes that we can usually view the policy process as a series of chains of command where political leaders articulate a clear policy preference which is then carried out at increasing levels of specificity as it goes through the administrative machinery that serves the government."[28] The top-down approach underlies the assumption that it is the responsibility of state institutions to provide immediate assistance that is relevant and coherent on the level of objectives and orientations.[29]

There is appreciation that policy design should be reduced to government decisions, focusing on the extent to which administrators carry out or fail to carry out the decisions.[30]

Policy formulation is then theorized in terms of knowledge about institutions of government, including detailed empirical examination of legislatures, courts, and bureaucracies, while generally ignoring the normative aspects of these institutions. An example of a disaster management plan designed to meet top-down principles is the Czech Republic flood control project initiated after the floods in 1997, and again in 2002. During 1997, the countries in Central and Eastern Europe were struck by heavy flooding that caused the deaths of 105–115 people and property damage estimated at $30 million.[31] Five years later, the region suffered heavy flooding but experienced fewer deaths and less property damage. This could be partly explained by the improved disaster plan that was initiated after the flood of 1997. The disaster emergency plan included an improved warning system, an efficient supply system of food and medical care, and improved coordination to facilitate evacuation efforts. The Czech Hydro-Meteorological Institute introduced various reforms, including improving its forecasting and early warning system to provide timely information to the general public. Czech Republic Emergency Medicine was instituted after 1997 to enhance medical supplies and the number of first aid–trained physicians. In late 2000, the Czech Republic enacted three general laws related to evacuation efforts to efficiently enforce evacuation orders and provide security and shelter to effected residents.[32]

Whatever the benefits, and there were many, these studies of the function of formal structures of political institutions in disaster management, for the most part remained descriptive, failing to generate the basis for evaluating the strengths, weaknesses, or purposes of such structures.

Thus, the major criticism of the top-down approach was that it offered little in the way of thinking about how policy problems should be approached, and it was virtually swept away by the concentration on senior decision-makers, who have a marginal role in making and implementing policies, compared to lower-level officials and private actors.

The contentless and contextless overtones of the top-down approach appealed enormously to policy makers as a clear-cut technical process that laid bare the essence of policy problems and the involvement of all private and public actors and institutions in the problem.

In sum, the most serious shortcomings of the top-down approach are as follows: The top-down approach demonstrates too narrow and static a picture of disaster management: policies are almost always related to "maintenance" and "protection," and addressed by measures from specific sector policies (nature conservation and agriculture). These typically lead to a functional division of policy making and implementation, which fails to acknowledge limitations on governments, thus, constraining the range of options they can choose to carry out the decisions. Internal and external constraints on government make public-policy making, and efforts to understand it, difficult indeed. The government's choice of a policy may be limited, for instance, due to shortage of resources, unclear understanding by implementers of the stated goals and activities, international and domestic pressure, or resistance to certain policy options. Thus, for example, we will not be able to achieve a comprehensive understanding of disaster management reforms across countries without recognizing powerful actors such as NGOs or grassroots organizations that are able to act against any government effort to maintain its centralization, to accelerate emergency activities.

Consequently, the top-down approach presents a lack of openness and responsiveness of policy decisions, objectives, and measures for public debate and bottom-up inputs. While participation should be enabled from the top down, and a large part of the responsibility for this lies with those in power (senior decision-makers), lower-level officials, administrators, private and public actors, and institutions also have crucial roles in administering these policy objectives and measures.

These shortcomings have been recognized by the bottom-up approach to management. This approach is based in the context of the longstanding mission of policy making as an agent of building social and state capacity, on the central importance of formal and informal interaction and communication between public and private actors and institutions, and those for whom the policy is intended. The policy making and implementing process reflects a more democratic structure that ensures that decisions made at the top include the interests of those at the bottom.[33]

The policy formulation process is conceived as strongly related to different multilevel governance structures. This involves a vertical governance that deals with the cooperation, coordination, and collaboration activities between local, subregional, regional, and national actors, and requires a bottom-up approach in order to address citizens' needs. Further, there is a double horizontal governance that deals with the cross-over among sectors (e.g., social inclusion, accessibility, spatial planning, and economic development), and the cross-over among different types of actors (i.e., public bodies, the associational sector, and the private sector).

Application of the bottom-up approach to management in the policy implementation process establishes a system of collective social responsibility shared by various actors involved in implementing programs. Thus, a broad sense of collective social responsibility that guarantees both the state and local communities equal responsibilities and advantages requires great transparency in the distribution of resources, guarded by effective monitoring and accountability systems. Disaster management programs that were developed in both Mozambique (2000 and 2001) and to some extent in Iran (2003) met the principles of the bottom-up approach. In 2000, Mozambique, one of the poorest countries in the world, faced heavy rainfall that led to massive flooding, causing the death of approximately 800 people and great property damage.[34] Following the disaster, the Mozambique National Contingency Plan was developed to enforce a coordinated network comprised of communities, districts, and provinces as well as local, national, and international agencies involved in emergency training to mitigate future disasters. Local authorities and NGOs provided special training programs to strengthen community leaders in running evacuation centers and in making use of local capacity, such as community-based social and medical service organizations.[35] The 2003 earthquake that struck Bam and the Kerman province of southeastern Iran caused the death of approximately 30,000 people and massive destruction.[36] The disaster response of the Iranian government is of special interest when viewed nowadays as, due to the earthquake, the relations between Iran and the United States thawed. The Iranian government had relied heavily on foreign assistance and on the local capacity of affected communities. The Iranian government coordinated with the United Nations and the Iran Red Crescent Society (IRCS) to mobilize local rescue teams that were already familiar with effected communities' needs, instead of instituting centralized bureaucracy.[37]

The latter exemplary practices in national disaster management reveal a shift of focus from a top-down approach to a bottom-up traditional approach in disaster management toward a mixed model in which the involvement of private and public actors in governance and management at the grassroots level becomes a vital component in disaster relief efforts. The spirit of this approach was admirably reflected by the Yokohama Strategy (1994), which addressed the guidelines outlined by the World Conference on Natural Disaster Reduction, held in Yokohama (Japan) in May 1994. The Yokohama strategy called for development of a "global culture of prevention" and improved risk assessment, broader monitoring and communication of warnings at the community and national levels, and at the regional and subregional levels. In 2000, the International Strategy of Disaster Reduction (ISDR) provided a framework to coordinate actions to address disaster risks at the local, national, regional, and international levels. It called for building resilient nations and communities as an essential condition for sustainable development. The Hyogo Framework for Action 2005–2015 (HFA), endorsed by 168 U.N. member states at the World Conference on Disaster Reduction in Kobe, Japan, in 2005, suggested a strategic and inclusive approach to reducing vulnerabilities and risk imposed by hazards. Among the principles for implementing disaster risk reduction guiding this ten-year plan are as follows:[38]

- Effective disaster risk reduction relies on the efforts of many different stakeholders, including regional and international organizations, civil society including volunteers, the private sector, the media, and the scientific community.
- A multihazard approach involves translating and linking knowledge of the full range of hazards into disaster and risk management, political strategies, professional assessments and technical analysis, and operational capabilities and public understanding, leading to greater effectiveness and cost efficiency.
- Capacity-development is a central strategy for reducing disaster risk. Capacity development is needed to build and maintain the ability of people, organizations, and societies to successfully manage their risks themselves. This requires not only training and specialized technical assistance, but also strengthening of the capacities of communities and individuals to recognize and reduce risks in their localities.

- Decentralization of responsibility is crucial for disaster risk reduction. In order to recognize and respond to these locally specific characteristics, it is necessary to decentralize responsibilities and resources for disaster risk reduction to relevant subnational or local authorities, as appropriate. Decentralization can also motivate increased local participation along with improved efficiency and equitable benefits from local services.
- Effective disaster risk reduction requires community participation. The involvement of communities in the design and implementation of activities helps to ensure that they are well tailored to the actual vulnerabilities and to the needs of the affected people.
- Public-private partnerships are an important tool for disaster risk reduction. Public-private partnerships are voluntary joint associations formed to address shared objectives through collaborative actions. They may involve public organizations such as government agencies, professional and/or academic institutions, and NGOs, together with business organizations such as companies, industry associations, and private foundations.

Following our brief review of international exemplary practices, it is argued that despite the attention devoted to the bottom-up approach in disaster management, the top-down approach appears to remain side-by-side with the bottom-up approach. These are not competing approaches but alternatives in disaster management policy analysis. Thus, to strengthen the synthesis of both approaches we need to comprehend the way in which existing disaster management practices meet the category of disaster vulnerability. Thus, the remainder of the book will attempt to assess the extent to which disaster management practices address the key components of disaster vulnerability. For that purpose, we need first to reflect on the conceptualization of vulnerability to generate evaluative criteria for assessing the appropriateness of disaster management practices to improve public administration disaster management performance.

2
Defining Disaster Vulnerability

The aim of this chapter is to explore the various definitions of *vulnerability* employed in food and human security analyses and to discuss them in light of the conceptualization of vulnerability appropriate for disaster management. This overview provides compelling reason to believe that vulnerability should be transformed into a more meaningful and usable construct for our understanding of the social justice issues surrounding disaster management policies.

Conceptualizing Vulnerability

A forerunner school of thought on natural hazards and disaster studies was pioneered by Barrows (1923) and White (1945),[1] who addressed the geographical approach to conceptualize human vulnerability. The investigation of the concept of vulnerability during the 1970s brought with it a paradigmatic shift in how scientists began to view human ecological adaptation to the environment, centering on individuals' behavior, such as their perceptions, attitudes, beliefs, values, response, and personalities. The studies on vulnerability within the context of natural disasters and hazards have followed the "spatiotemporal" distribution of hazard impacts on vulnerability, and people's choice and adjustment to the natural hazards approach of the Chicago-Colorado-Clark-Toronto School of Natural Hazard Studies mostly associated with Kates, White, and Burton et al.[2]

This shift of focus from technical vulnerability based on engineering and natural science, which delimited the role of human systems in the assessment of hazards and their impacts, to socioeconomic vulnerability underlying the role of human socioeconomic indicators in mediating hazards' consequences, was widely applied in social science fields such as sociology and economics.[3] The concept of vulnerability within the context of natural disasters has been considered in relation to specific hazards or causes due to temporal and spatial patterns of the events, such as food security and human security studies. In both fields, vulnerability was defined as society's exposure to external hazards and their socioeconomically determined coping capacity, including inflation rate, density, age, and illiteracy rate.[4] Mitigating the effects of natural disasters and hazards would therefore involve reducing vulnerability through socioeconomic interventions, for "a society's pattern of vulnerability is a core element of a disaster. It conditions the behavior of individuals and organizations throughout the full unfolding of a disaster far more profoundly than will the physical force of the destructive agent."[5] However, there is no common definition of vulnerability in the social sciences.[6]

In the first part of this chapter, we draw on a variety of conceptualizations of vulnerability theory to describe the constitutive features that define vulnerability in food and human security literatures and the potential effects of its use as a tool for measuring vulnerability to food and human insecurity. We have chosen two attributes that are considered by most scholars as central components to the understanding of vulnerability: risk factors (or vulnerability causes) and capacity. Following this brief survey, we offer a definition of vulnerability that briefly illustrates the application of the concept by describing some of the principal ways in which natural disaster increases risk and erodes coping mechanisms. The discussion concludes with suggestions for transforming vulnerability into a more meaningful and usable construct for assessing disaster management practices.

Vulnerability in Food Security

In food security, the first dimension of vulnerability revolves around the concept of capabilities, which refer to people's lack of capacity to define their own life choices and to pursue their own goals, even in the face of adversity. The capability approach argues that evaluating the status of individuals and societies must go beyond income, utility, rights, and resources to the actual lives we lead: the freedoms and opportunities to be and to do that people have reason to value, their quality of life.[7] Assessment of quality of life is measured in terms of working conditions, access to leisure, and degree of social integration

According to Sen,[8] capabilities are rooted in the potential that people have for living the lives they want, of achieving valued ways of "being and doing." Sen's theory addresses the idea of "functionings" to refer to all possible ways of access in a given context, and of "functioning achievements" to refer to the particular ways of being and doing that are realized by different individuals. Clearly, the failure to achieve valued ways of "being and doing" can be traced to vulnerability. It is only when the failure to achieve one's goals reflects some deep-seated constraint on the ability to choose that it can be taken as a manifestation of vulnerability. In the domain of food security, Sen's work on famine points to the fact that famine could result from a failure in people's ability to have access to food.[9] Vulnerability analysis is based on the identification of individuals and households rather than whole populations who are unable to obtain food, and the socioeconomic variables that restrict their access to food.

However, commentators have criticized Sen for his failure to be specific about which capabilities matter, which ones more than others, and why.[10] In addition, Sen's failure to articulate and justify a theory of social justice is the shortcoming of his work.[11]

The studies that followed Sen's have raised concerns regarding the variability of risk exposure in achieving optimal functioning. It is argued that individuals exposed to particular adverse life circumstances are treated as homogenous groups, despite the possible variation in the degree to which they are influenced by other risk factors in question, without more comprehensive knowledge about risk factors that are involved in determining vulnerability. Exposure is not simply about exposure to a temporary state of famine; it is more than just a crisis in overall food supply, but can be a long-term and gradual process emanating from the accumulation of risk effects. Thus, to avoid an overly narrow definition of vulnerability, Chambers[12] proposed a double structure model of vulnerability.

According to Chambers, vulnerability should defined as an "exposure to contingencies and stresses and the difficulty which some communities experience while coping with such contingencies and stresses."[13] Following this assumption, vulnerability as a twofold term incorporates an external dimension, which addresses the exposure to external shocks and stresses, and an internal dimension, which relates to incapacity to cope. Although this model of vulnerability encompasses a multiple range of risk exposure, the "internal" dimension of vulnerability has been less well understood, as the ability to mitigate risks is a highly complex, contextual, and dynamic issue, especially in times of acute crisis, since coping is also shaped by daily or socioeconomic risk factors threatening not only current well-being or functioning but also longer-term security.[14] Moreover, the internal side of vulnerability is highly context-specific and is often not visible.

In addition, it is argued that the external dimension of vulnerability can also be shaped by external shocks and stresses that transcend household and national boundaries, such as economic globalization, urbanization, infectious diseases, political conflicts, and environmental challenges.[15]

Although splitting vulnerability into two dimensions can contribute to the understanding of risk factors that shape vulnerability, little is known about the intersections and interactions between these dimensions, and of the different levels in the broader systems within which households are embedded. Therefore, identification of the variability of risk exposure as influenced by the interplay of multiple factors and across multiple spheres is relative, situational, and attributional, whereby identified factors such as income, gender, age, disability, and location may have greater significance to the researchers who define vulnerability than the individuals who actually experience it.[16] Ellis, for example, demonstrates the concern addressed by the plurality of meaning or interpretations in evaluation of risk exposure by local people who diversify their capabilities according to their perceptions of risk and in terms of available risk management strategies.[17] For that, coping with stressors may involve a reactive or anticipatory strategy.

Stephen Devereux has demonstrated that there is also diversity in the temporal dimension of vulnerability in the domain of food security, much of which can be only short term (temporary or transitory), but other vulnerabilities can be for a long duration (chronic).[18] Early experience of risk exposure may be overcome by improved conditions, but may nevertheless leave households potentially more vulnerable to disadvantage or other risks experienced at a later stage. This is indicated by Devereux, as "People who tend to be vulnerable to food crises are often chronically (but moderately) food insecure, even before a livelihood shock raises their food insecurity ('transitorily') to life-threatening levels."[19] However, the temporal dimension alone cannot determine food insecurity but should be combined with the level of the severity of food insecurity (moderate or acute, for example) to denote the existence of chronic insecurity, and moderate or severe transitory insecurity forms of vulnerability.[20] He puts forward a new concept, "composite food insecurity," to denote the way in which "households that are chronically 'hungry' at the best of times, . . . are also susceptible to periodic food shocks," thus reflecting the dynamic shift in household-environment interactions between moderate and severe insecurity.[21] Devereux stresses the importance of the dynamic characteristic of household-environment interactions in the face of risk exposure, which brings new impetus to the design and development of policy interventions:

> In situations of "severe transitory" food insecurity requiring an immediate humanitarian response, interventions must be based not only on an understanding of what caused the food shock, but also on a disaggregated understanding of the impacts of the food shock on individuals and households with differential abilities to cope.[22]

Watts and Bohle[23] and Bohle[24] further developed the model introduced by Chambers[25] by including indicators of recovery potential in their analysis of food supply. A set of qualifications refers to the conditions of vulnerability, including exposure to risk, coping capacity, and recovery potential. Thus, the causes of vulnerability are evaluated in terms of their transformatory significance; the extent to which assessment of recovery potential has the potential for challenging and destabilizing social inequalities rather than reproducing such inequalities. Thus such conceptualization of vulnerability has also highlighted the interdependence of individual and structural change in processes of coping with adversity. Access to assets can tell us about potential rather than actual recovery ability, and the validity of an asset measure as an indicator of vulnerability largely rests on the validity of the assumptions made about the potential entitlement embodied in that asset. Thus, considerations apply to evidence on vulnerability: we have to know about its consequential significance in terms of individuals' recovery and the extent to which it had transformatory potential. The more evidence there is to support these assumptions, the more confidence we are likely to have in the validity of the indicator in question. As seen, indicators of vulnerability cannot provide an accurate measurement of individuals' or households' ability to cope with stresses of food supply, they merely have to indicate the direction and meaning of vulnerability. Although considering assets and entitlements, they are unlikely to automatically result in coping and recovery, but they do create the vantage point of alternatives, which allows a more transformatory consciousness to come into play. The translation of these assets and opportunities into the kinds of functioning achievements such as recovery potential, which would signal coping capacity, is likely to be closely influenced by the possibilities for transformation on the ground, and how they are perceived and assessed.

This conceptualization of vulnerability has been adopted by researchers in the domain of natural disasters. For example, Blaikie et al.[26] have used this model by applying the degree to which individuals or groups can recover from exposure to the effects of a natural hazard that goes beyond those found in the food supply. As formulated by Blaikie et al., "social systems ... generate disasters by making people vulnerable."[27] Blaikie et al. raise the role of inherent institutional classism and inequality in intensifying the sense of social and environmental vulnerability, which is likely to lead to a sense of powerlessness and inability to exercise individual or collective will when facing natural disaster.

Frank Riely[28] integrates exposure and coping capacity into the risk equation and, specifically, into the vulnerability components advanced by Chambers's model as both a function of exposure to shocks (or risk factors) as well as "underlying socioeconomic processes which serve to reduce the capacity of populations to cope with those risks."[29] These two components of vulnerability are summarized by Riely's vulnerability equation as:

Vulnerability = Exposure to Risk + Ability to Cope.

To conceptualize vulnerability in terms of assets and recovery potential bears the advantage of treating vulnerability as a dynamic process, rather than a state or trait. Indeed, by viewing coping capacity or functioning in the face of limited food supply is not only dependent on the characteristics of the individual, but is greatly influenced by processes and interactions arising in different levels and factors of risk exposure. Yet, in bringing issues of variations of individual response only the general relation between parts of the progression of vulnerability is provided.

Vulnerability, as used in food security literature, is defined as people's and households' exposure to external hazards and their socioeconomically determined capacity to cope in relation to hunger, food insecurity, or famine. The focus of most of the studies on vulnerability in food security has been on the identification of vulnerable groups within a population, rather than a specific hazard. Given the central importance to determining where a particular group of people is positioned along a relative scale of vulnerability, research findings have illustrated a variety of causes of the vulnerability of specific groups. In this way, "Ambiguity with respect to causality and non-contingent, a priori vulnerability lead to a tendency to formulate vulnerability indices incorporating every conceivable available variable—from rainfall and vegetative vigor to health and nutrition data."[30]

From a policy perspective, Dilley and Boudreau suggest a shift from pathologizing individuals and households at risk;[31] the study on vulnerability needs to provide an overarching framework for conceptualizing causes of their vulnerability, intervention strategies, and practice. Following this criticism, of central interest are not causes or incapacities, which are the traditional focus of research on high-risk groups, but positive coping outcomes (e.g., recovery) and their antecedents. However, from a distributive perspective, such a reactive approach may lead to a reduction of what is spent on disaster prevention and mitigation.

Despite the critique of placing exclusive attention on identification of groups at risk and the causes of vulnerability leading to greater abstraction in assessing vulnerability in food security, this conceptualization was applied in developing the social vulnerability index (SVI) in emergency and disaster management literature. In 2011, Flanagan et al. issued a social vulnerability index incorporating fifteen census variables across various fields, including socioeconomic status comprising income, poverty, employment, and education variables; household composition; disability comprising age, single parenting, and disability variables; minority status; language comprising race, ethnicity, and English language proficiency variables; and housing and transportation comprising housing structure, crowding, and vehicle access variables.[32] The index aims at assisting state, local, and tribal disaster management officials to identify the locations of their most vulnerable populations.

Vulnerability in Human Security

The concept of vulnerability was introduced in human security studies to capture the impact of globalization threats to human livelihood. The impact of globalization on society and on human livelihoods is by and large conducted in terms of familiar conceptual categories such as human security.[33] Introducing the concept of vulnerability draws attention to identification of the wide range of threats to well-being imposed by globalization, including "physical insecurity, threats to state autonomy, instability, and vulnerability which is described as 'a susceptibility to damage.'"[34]

When employed in the determination of human security in this way, the focus of vulnerability has shifted from right-based conceptualization encompassing a humanitarian and liberal conception of "life, liberty, and the pursuit of happiness"[35] to the capacity-based approach that dominated food security studies.

In addressing the distinctive aspects of vulnerability generated by the impact of globalization on society since the 1990s, Joseph Stiglitz maintains that: "Even many of those who are better off feel more vulnerable."[36] This claim underlies the need to embed vulnerability in a theoretical framework that refers not only to the exposure of increased risks imposed by globalization, but to the growing incapacity to manage those risks in guiding the investigation and understanding that shape society's vulnerability. To this extent, globalization is often viewed as a double-edged sword: individuals can choose between alternatives, competing goods and services, on one hand, but may also face barriers to achieving autonomy of decision making due to lack of access to information on risks and consequences.

Thus, exposure to globalization threats rests on the two dimensions of vulnerability addressed in food security literature: the ability of individuals, communities, or states to cope with the effects of economic globalization, urbanization, infectious diseases, military and political conflicts, and environmental changes; and the consideration of vulnerability causes. The centrality of both dimensions is articulated by the United Nations Department of Economic and Social Affairs, which offers the following definition:

> In essence, vulnerability can be seen as a state of high exposure to certain risks and uncertainties, in combination with a reduced ability to protect or defend oneself against those risks and uncertainties and cope with their negative consequences. It exists at all levels and dimensions of society and forms an integral part of the human condition, affecting both individuals and society as a whole.[37]

It is then worth showing how these features have been addressed by intergovernmental organizations in discussing vulnerability at the global level. During the end of the 1990s, several IGOs (intergovernmental organizations) began to address the concept of vulnerability to present distribution and inequality of power and assets. The U.N. Department of Economic and Social Affairs' report on the World Social Situation in 2003 applied the concept of vulnerability.[38] According to the report:

> While vulnerability, uncertainty and insecurity in the life of people are not new, what is new is that their causes and manifestations have multiplied and changed profoundly over the last decade. Examples include civil strife and the proliferation of conflicts, growing inequalities within and among countries further accentuated by globalization, mixed outcomes of poverty reduction efforts, increased mobility of populations and changes in family structures.[39]

Central to the discussion on vulnerability within the global context is the identification of at-risk groups that are especially vulnerable, including the young, the elderly, people with disabilities, migrants, and indigenous peoples. This report provides social and economic aspects of vulnerability: "All groups face vulnerabilities that are largely the outcome of economic, social and cultural barriers that restrict opportunities for, and impede the social integration/participation of the groups."[40]

Early on, the U.N. Development Programme (UNDP) addressed a similar comprehensive view of vulnerability in identifying vulnerable groups, since "people everywhere are more vulnerable":[41]

Changing labor markets are making people insecure in their jobs and livelihoods. The erosion of the welfare state removes safety nets. And the financial crisis is now a social crisis. All this is happening as globalization erodes the fiscal base of countries, particularly developing countries, shrinking the public resources and institutions to protect people.

Scholars questioned whether the welfare state as a modern economic mechanism has sufficient capacity to cope with increased risks. As argued by Mishra, "True, many European nationals have inherited a large welfare state from the golden age and, for the moment, seem to be able to hold on to them. But can they hold out against global pressures?"[42]

Such conceptualization was also taken up by the U.N. Economic Commission for Latin America and the Caribbean (ECLAC). The identification of vulnerable target groups rests on relative, situational, and perceptional conceptualization of risk, assets, and the ability to cope "to such an extent that vulnerability may be regarded as a distinctive feature of the social situation in the 1990s."[43] Thus, the threats imposed by globalization situate vulnerability as a distinctive problem, since "The opening up of markets and the downgrading of the state's role in the economy and society have exacerbated the insecurity and defenselessness affecting large groups of individuals and families, who are now exposed to increased risk."[44]

Increased environmental risks are also becoming threats to human security and being integrated in the conceptualization of vulnerability causes. In 1992, the International Panel of Climate Change (IPCC) defined vulnerability in terms of inability to cope with the consequences of climate change and sea-level rise. This definition was also pursued by Watson et al. "as the extent to which climate change may damage or harm a system; it depends not only on a system's sensitivity but also on its ability to adapt to new climatic conditions."[45] The U.N. Environmental Programme (UNEP) acknowledges environmental vulnerability as an essential element of human insecurity, "Since everyone is vulnerable to environmental threats, in some way, the issue cuts across rich and poor, urban and rural, North and South, and may undermine the entire sustainable development process in developing . . . coping capacity that was adequate in the past has not kept pace with environmental change."[46]

Conversely, such variability introduced to the conceptualization of vulnerability in identifying vulnerable groups proves more difficult in operationalization and measurement of the vulnerability of individuals and collectives. One of the initiatives to recognize the need to provide measurement of vulnerability was carried out by the U.N. Economic and Social Council in 2000. The council developed an Economic Vulnerability Index (EVI) based on five indictors to measure countries' economic vulnerability, including the magnitude of external shocks beyond domestic control, the exposure of the economy to these shocks, and the structural barriers explaining the country's high exposure.[47] A similar index was offered earlier by the International Monetary Fund (IMF) in developing vulnerability indicators to identify countries that are vulnerable to financial currency crises, which then occurred in Mexico, East Asia, Russia, and Brazil. The Fund states that "Timely and detailed data on international reserves, external debt, and capital flows strengthen the ability to detect vulnerabilities, giving policy makers enough time to put remedial measures in place."[48] The World Bank, for example, has offered vulnerability indicators to identify those states most vulnerable only in terms of the globalized economy.[49]

However, it should be indicated that on the regional level, developing a vulnerability index has shown greater effort to go beyond economic variables. For instance, in the Caribbean region, ECLAC has developed a Social Vulnerability Index (SVI) comprised from various indicators such as poverty, crime, natural disasters, migration, health status, and social marginalization.[50]

A focus on the development of vulnerability indicators to construct an index of vulnerability bears the danger of identifying vulnerability in terms of the characteristics of states or groups who are likely to be exposed to increased risks, following the assumption that states or groups are rendered responsible for their vulnerability. Factors such as economic growth, employment, social stability, and political power have been used to characterize strong states that, due to these attributes, are able to face stress without permanent damage.[51] Yet, although states may manifest vulnerability in their actions, vulnerability is not a static characteristic. Coping capacity and functioning in the face of risk exposure is not only dependent on the states' characteristics but is greatly influenced by processes and interactions arising from the state and global contexts. Coping capacity is continually produced and changed by the socioeconomic context experienced by vulnerable groups. A step in incorporating coping capacity in the index of vulnerability is provided by the hazards-of-place model of vulnerability. The model assesses the hazard potential or the overall place vulnerability as resulting from the way a geographic place interacts with the social fabric of the place. The social fabric refers to community coping experience with hazards, which in turn is influenced by economic, demographic, and housing characteristics.[52] What is vital for a better understanding of vulnerability is the consideration of dynamic group-environment interactions reflecting coping capacity to mitigate risks. This has been recognized by Keohane and Nye, who conceptualized vulnerability by relating it to "complex interdependence" understanding of the social impact of globalization: "The vulnerability dimension of interdependence rests on the relative availability and costliness of the alternatives that various actors face."[53] What, therefore, distinguishes vulnerability conceptualization from other conceptualizations widely used to analyze vulnerability in human security? It is argued that the utilization of choice between alternatives offers a far more optimistic assessment for action aiming to promote coping capacity and reduce vulnerability, both in terms of its analytical reach and in terms of what it points to by way of policy responses (however, it largely limits itself to the action of states or intergovernmental bodies). This conceptualization of vulnerability is manifested by ECLAC as: social vulnerability is "the net effect of the competition between social risks and social resilience" where it views resilience as "tantamount to an ability that is based on entitlement, enfranchisement, empowerment and capabilities."[54]

According to Kirby,[55] such conceptualization can be traced back to the analytical framework provided by Karl Polanyi.[56] The term *vulnerability* constitutes an essential part in Polanyi's explanation of how and in what way the market comes to determine more and more of how we think and what we value, of what is produced and by whom, of how it is distributed, and of how all these affect society, livelihoods, and quality of life. Viewed in this way, it is suggested by Polani that the vulnerability is generated from the way the market mechanism runs society. Following the assumption that the human person is not an economic being "but a social being," human security lies in the sense of belonging to society, which is ignored by the market society, "since market situations do not, in principle, know wants and needs other than those expressed by individuals, and wants and needs are here restricted to things that can be supplied in a market, any discussion of the nature of human wants and needs in general was without substance."[57] Vulnerability cannot be defined by access to market entitlement that utilizes a narrowing conceptualization based only on motivations that lead to the destruction of social bonds and sentiments of social belonging on which the human person's well-being depends, for "We are faced with the vital task of restoring the fullness of life to the person, even though this may mean a technologically less efficient society."[58]

As emphasized in his discussion in *The Great Transformation* of the debates about the social impact of the Industrial Revolution, economic vulnerability should be associated with sudden social dislocation or inability to maintain social ties rather than with low income. According to Polanyi, "Not economic exploitation, as often assumed, but the disintegration of the cultural environment of the victim is then the cause of the degradation." The essence of this disintegration lies in the lethal injury to the institutions in which his social existence is embodied. "Loss of self-respect and standards, whether the unit is a people or a class, whether the process springs from so-called 'culture conflict' or from a change in the position of a class within the confines of a society."[59]

Polanyi's work offers an exceptional conceptualization of vulnerability that bears the potential to achieve an analytical precision that most definitions cannot claim. Drawing on insights from Polanyi's discussion of vulnerability, it is argued that vulnerability should be treated as a relational concept grounded in the roots of social order.[60] As a relational conception of human security, vulnerability underlies the sentiment of belonging to society rather than individualistic understanding. In this way, vulnerability as a relational and collectivist phenomenon depends on complex interactions between human groups and the environment that bring about the actions and means that strengthen social bonds and help to cope with the threats to which collectivity is vulnerable. Thus, the measurement of vulnerability by various indicators associated with globalization as developed by IGOs' initiatives should evaluate vulnerability in terms of its transformatory (dynamic) significance, that is, the extent to which the risks assessments made have the potential to challenge and destabilize social inequalities rather than express and reproduce those inequalities.[61] Hence, for vulnerability to be a useful concept for analysis, it must be embedded in a normative framework that guides appropriate policy interventions to manage risks.

Bringing the value of social belonging rather than the individual's well-being to the attention of policy makers and disaster management shifts the focus on social assets that provide support networks such as the family, local neighborhood associational life, and communities. Despite the fact that such collectives find themselves strained in today's world, they still have a greater role in the provision of caring services and in mitigating the risks imposed by natural disasters.

We began by noting that there is considerable concern about, and discussion of, vulnerability in food and human security contexts and its normative significance. Thus, there is a debate about whether disaster vulnerability deserves an exceptional treatment by way of regulation. Methodologically, the review of attempts to measure coping capacity has been about the various indicators to identify vulnerability on the level of regions and nations. However, those indicators of vulnerability revealed empirically cannot provide an accurate measurement of changes in coping capacity, they merely have to identify at-risk groups and a general direction of change. Social indicators of vulnerability can lend themselves to a variety of different and contradictory meanings that may not operate in the same manner for all groups, all contexts, or all outcomes. For that, evaluation of institutions for disaster management requires more comprehensive and discriminatory assessments of distributional consequences; more powerful criteria must be sought. What is required are value-laden, collective, and dynamic-based interventions and integrated service delivery. Thus, in the following chapter we evaluate how well-established disaster management doctrines meet vulnerability criteria as reflected in the brief overview presented in this chapter. This includes the following:

- **Multidimensional risk categories**: Vulnerability includes a multidimensional range of risks. Any assessment at the national level must take into account economic, social, cultural, political, and environmental patterns of vulnerability within the distributive arrangements of a country.
- **Dynamic and Interactive processes**: Vulnerability is viewed as a dynamic process, subject to constant evolution of a complex of interactive processes affecting the bonds that constitute society and its coping capacity.

- **Collective and Shared interests**: Vulnerability is viewed within a broader sense of community and societal interests and responsibility.
- **Procedural/Structural constraints**: Possibilities and alternatives in reducing vulnerability relate to the structural conditions under which responses are made. These constraints may inhibit the choice of some policy options and promote others. For that, political and administrative interventions should strengthen coping mechanisms and the availability of appropriate information.
- **Integration of normative evaluation**: An integrated disaster management should employ the integration of administrative mechanisms as well as appropriate ethical and professional orientations, leading to increased capabilities of staff responding to communities with multiple needs. Without an ethical-normative foundation, neither individual actors nor agencies perform their proper function.

3
Vulnerability Assessment of Disaster Management Doctrines

Drawing on the conceptualization of vulnerability presented in the previous chapter, the role norms and values play in assessing vulnerability has been largely overlooked.

Norms and social-cultural values have an influence over how citizens evaluate the distribution of needed goods and services to reduce vulnerability. In disaster management, normative considerations play an important role in service delivery.

The need to apply normative evaluation to disaster management is recognized by a number of studies. For example, McEntire et al. and others have developed a Comprehensive Vulnerability Management model for disaster emergency management (DEM).[1] The model attempts to treat every disaster as a shared responsibility, taking into account a wide array of actors, triggering agents, functional areas, and normative assessment of capacity building.

Such a framework concentrates on how vulnerability is constructed by economic, social, institutional, cultural, political, and psychological variables as well as capacity building.[2] This framework has three key related principles, listed below.

First, the Comprehensive Vulnerability Management framework emphasizes the role of the public sector to reduce disaster liabilities and increase capacity building.[3]

Second, the Comprehensive Vulnerability Management framework concentrates on various triggering agents that interact with each other in both preparedness and planning stages, as well as in recovery and local capacity-building areas of emergency management.[4]

Third, the Comprehensive Vulnerability Management framework

provides multidimensional categorization of vulnerability, including economic, social, institutional, cultural, political, and psychological variables used to assess capacity building.

Following these principles, the Comprehensive Vulnerability Management strategy offers a holistic model to broaden the scope of emergency management concerns, as reflected in the "whole community" framework recently applied by the Federal Emergency Management Agency (FEMA) in the United States.[5]

This book aims to complement their work by examining how well the disaster management doctrines on which disaster management policies are based conform to social justice criteria. In evaluating the disaster management doctrines, this chapter first questions whether the principles underpinning the doctrines of resistance, sustainability, and resilience meet vulnerability criteria set by the Comprehensive Vulnerability Management model. Having presented the evaluation of three disaster management doctrines, this chapter explores which doctrine has the potential for being fully consistent with norms of social justice and fairness.

Vulnerability-based disaster management: From Resistance to Comprehensive Vulnerability Doctrines

Resistance Doctrine

The potential for resistance appears to be an important variable for counseling and clinical practice.[6] Resistance has been acknowledged as a general tendency of subjects to defend themselves in response to a threat in both counseling and clinical practice in which patients either directly or indirectly are reluctant to change their behavior or refuse to share, remember, or recognize feelings and motives of presumably clinically relevant experiences.[7] Resistance originated in Freud's theory of psychoanalysis to denote a process:

> The discovery of the unconscious and the introduction of it into consciousness is performed in the face of a continuous resistance (*Widerstände*) on the part of the patient. The process of bringing this unconscious material to light is associated with pain (*Unlust*), and because of this pain the patient again and again

rejects it . . . It is for you then to interpose in this conflict in the patient's mental life. If you succeed in persuading him to accept, by virtue of a better understanding, something that up to now, in consequence of this automatic regulation by pain, he has rejected (repressed), you will then have accomplished something towards his education . . . Psychoanalytic treatment may in general be conceived of as such a re-education in overcoming internal resistances.[8]

In this regard, Freud posited that resistance is manifested through primary gain (internal benefits) and secondary gain (external benefits) derived from patients' psychiatric symptoms.

In civil life illness can be used as a screen to gloss over incompetence in one's profession or in competition with other people; while in the family it can serve as a means for sacrificing the other members and extorting proofs of their love or for imposing one's will upon them . . . we sum it up in the term "gain from illness" . . . But there are other motives, that lie still deeper, for holding on to being ill . . . [b]ut these cannot be understood without a fresh journey into psychological theory.[9]

Thus, resistance is discovered in those particularly sensitive areas of the patient's history that are protected by layers of defense mechanisms.[10]

The goal of treating manifestations of resistance is to discover the subject's recollections of particularly painful events from which the client sought emotional protection, resistance through behaviors such as avoidance, and yet expressing unacceptable drives, feelings, fantasies, and negative behavior patterns.[11] The underlying assumption in responding to a subject's resistance is that the subject's symptoms and problematic behaviors should be seen as survival mechanisms. This means that there is a need to respect resistance and various forms of avoidance, and to strive to understand the meaning of such behaviors. Part of respecting resistance means empathizing with the subject and his reason for feeling resistant.

Mitigating risks in terms of resistance strategy bears the need to strengthen local communities and their resources to "represent the safest possible community that we have the knowledge to design and build in a natural hazard context" in a way that "the ability of structures and infrastructures to withstand the forces of powerful agents, minimize[s] damages."[12]

Building a disaster-resistant community should be based on three dimensions, including the natural risk factor, appropriate sector (urban planning and engineering), and institutional capacity, to effectively implement emergency plans and arrange the roles of communities, private actors, local governments, and professionals. Thus, dealing effectively with defensive behavior policy instruments should cover each of the functional areas of emergency management. Policy instruments that meet the stated goals of resistance strategy are targeted to strengthen safety by utilizing construction practices, building codes, and technology that reduce property and life losses from known natural hazards.[13] An example of the application of resistance policy instruments is the disaster management plan that was developed in Cuba, which was struck by ten hurricanes between 1995 and 2005.[14] Such devastating experience has led Cuba to predesign houses and shelters as hurricane safe, stocked with food, water, medical, and other emergency supplies, and able to accommodate the large number of evacuees. The evacuation preparedness procedures were well coordinated by one central agency—the National Civil Defense. In addition, Cuba preestablished medical teams to offer immediate medical assistance when a disaster strikes.[15]

The advantages associated with the disaster-resistant community strategy lie in the encouragement of communities to participate in mitigation processes and the use of a spatially direct set of technological and economic measures to provide safety and decrease the degree of loss resulting from physical agents, which is technically fairly easy to accomplish: "Provide the best means for developing the most effective disaster and emergency management programs."[16] In his study, Aguirre[17] criticized the success of the disaster program applied in Cuba. According to Aguirre, when viewed from a narrow resistant perspective, it could be said that Cuba performed an effective disaster preparedness and response involving warning and evacuation; however, when evaluating Cuba's disaster management performance from a normative perspective, it is claimed that socioeconomic and political issues that influence the adaptive capacity of the population as a whole have been ignored.

Drawing on the evaluative criteria of vulnerability, it is argued that resistance strategy *overlooks normative aspects* that are involved in promoting the well-being of communities at risk, such as a sense of collectivity including community ties, value of solidarity, shared interests and values, and common goals. This strategy relies heavily on technical rather than ethical-professional performance of actors and institutions that are engaged in mitigation planning such as social workers, psychologists, educators, and nongovernmental organizations.

Moreover, resistance strategy views a disaster-resistant community as an outcome rather than an interactive process whose continuous nature makes it difficult for public administrators to coordinate with local, nongovernmental, and private organizations, due to the constraints of the strong command and control structure needed to enhance resistance to hazards and emergency dynamics.[18] In the short run, with natural disasters there is certainty that resistant communities will face some temporary alteration of economic activity and less physical impact within the affected area. However, natural disasters do affect the interrelationship of community members, households, and local businesses. Thus, resistance strategy may have greater effect on infrastructures and economic development than on long-run community capacities assessed by social science disciplines such as public administration, sociology, economics, political science, anthropology, and psychology.

Sustainability Doctrine

Another strategy to address the vulnerability of communities and their capacities is sustainability. Despite its use in multiple domains, there is a growing uncertainty regarding the meaning of sustainability in the academic literature on environmental and social values, politics, and economics. In this section we suggest a more coherent conceptualization of sustainability that will allow us to engage the question of how sustainability framework is applied to disaster management in pursuit of a bottom-up approach.

A suggested route into the discussions surrounding the conceptualization of sustainability has been to trace its roots in environmental and developmental thinking. The concept of sustainability was first adopted by the United Nations Conference on Human Environment held in Stockholm in 1972. During the 1980s, it became a key concept in the *World Conservation Strategy*[19] and in the *Brundtland Commission report*, issued by the World Council on

Economic Development, which laid the groundwork for convening the Earth Summit in Rio de Janeiro five years later. Held in June 1992, the Rio Summit was the largest environmental conference ever organized, bringing together over 30,000 participants, including more than 100 heads of state. The World Commission on Environment and Development referred to sustainability as "development that meets the needs of the present without compromising the ability of future generations to meet their own needs."[20] The report suggested that sustainability should follow specific objectives, including ensuring grassroots involvement in decision making for conserving and enhancing natural resources and adopting appropriate technologies (smaller-scale power plants, energy conservation).

Thus, the Brundtland Commission report paved the way for viewing sustainability from a bottom-up approach in preservation of natural resources. This notion of sustainability, borrowed from ecological studies of nonhuman populations, as applied to disaster emergency management plays a large integrative role. Beatley suggested viewing sustainability in a natural disaster setting as "where people and property are kept out the way of natural hazards, where the inherently mitigating qualities of natural environmental systems are maintained, and where development is maintained."[21] In a study on coastal storms, Godschalk, Brower, and Beatley identified the importance of incorporating physical factors as well as social and environmental forces such as social capital, political support, and opposition from various stakeholders concerned with designing infrastructure policies into sustainable emergency programs.[22] Mileti's research, which encompasses a large variety of natural disasters, suggested that "A locality can tolerate—and overcome—damage, diminished productivity, and reduced quality of life from an extreme event without significant outside assistance."[23]

Mileti linked natural disasters to the sociocultural tradition by emphasizing maintenance of the community as well as of social values such as equity and quality of life. He maintained that neither resources nor economic welfare are sufficient as objectives of sustainability if the social community and its social values are not also pursued. He also emphasized the importance of institutions and institutional capacity in implementing mitigating policies for the long run. Mileti offered the following five policy instruments appropriate to achieving sustainable hazards mitigation:

1. better land-use planning and management to limit settlement in dangerous areas;
2. enforcement of building codes and standards to protect people and property;
3. increased reliance on insurance to cover possible financial losses from disaster;
4. enhanced prediction, forecasting, and warning systems; and
5. improved engineering for buildings and infrastructure to minimize death and damage associated with disaster.[24]

The volcanic eruption relief efforts that took place in the Philippines in 1991 addressed the principles of sustainability strategy. Disaster management developed by government agencies as well as warning systems include advanced technological systems coupled with local warning reports. Other policy instruments included communication and educational instruments such as civic education and information campaigns designed for both the general public and at-risk communities.[25]

Mileti's model of sustainability prevails in disaster management literature since it incorporates a broad spectrum of risk factors and processes in disaster management and their impact on building community capacities in times of disaster.

Moreover, the sustainability concept encompasses outcomes and practices across domains to formulate linkages between hazard mitigation and disaster recovery. Unless multiple domains of disaster management are assessed, only a partial picture of sustainability can be established. Sustainability also meets societal expectations associated with vulnerability by drawing on variables such as culture, economics, and environment in coping with adversity.

The conceptualization of sustainability offered by Mileti is not without criticism, and serious concerns have been raised about the practical/procedural value of this strategy—about how it has been measured and how it has been applied in disaster mitigation practices. In particular, concerns have been identified with its limited capacity to deal with only extreme natural events and specific types of technological triggering agents.[26] A major limitation of the concept is that it is tied to extreme climate and tectonic events (earthquakes, volcanic eruptions). Although these events can have profound impacts on human life and natural resources, socioeconomic events, such as major price and market changes, and economic and political upheavals, gain little if any conceptual or empirical exploration in sustainable studies. In addition, sustainability does not render a system better able to coordinate between relevant actors. As Berke indicated:

> The interest groups involved in mitigation . . . and long-range disaster recovery are likely to be closely associated with the interests of sustainable development advocates. However, for those interest groups concerned with emergency preparedness and response issues (e.g., disaster warning, search and rescue, evacuation, and sheltering) the relationship with sustainable development would be less salient.[27]

One explanation for the relatively low level of integration of various disciplines such as education, sociology, political science, anthropology, psychology, and epidemiology, in addition to the functional areas, may be that sustainability management has consistently given the highest priority to concern for extreme natural events and variable conditions. From a policy perspective, intervention or prevention programs should be integrated into the cultural context, the educational programs, and the psychological and behavioral repertoire of vulnerable communities. Insights are gained from consultation with service users, psychologists, health-care practitioners, and so on. The notion that there is one basic mechanism underlying sustainability in different contexts and at different disaster management stages is untenable. Unless there will be a great diversity in pathways leading to sustainability, requiring both technical and sociocultural prevention strategies, sustainability will remain a normative relativist statement:

> It is clear that all adverse impacts of [disasters] will not be eliminated as is currently put forth in much of the sustainable development literature. The knowledge gained by [disaster] researchers and the extensive experience of [disaster] practitioners needs to be meaningfully introduced into the sustainable development debate. Otherwise, naive assumptions about sustainable development eliminating [disaster] impacts could lead to the shaping of flawed policy.[28]

Resilience Doctrine

In the past two decades, resilience—the ability to withstand and recover from adversity—has become a key concept in studies from a variety of disciplines. The disciplines of public health, sociology, psychology, and mental health, converged at a similar position, propose the following question: "What accounts for why some do well in the face of risk and adversity and others do not?" Competent functioning in the face of adversity has been defined as resilience. In these studies, the evidence for resilience was usually based on multiple domains (such as social relationships or employment or good health) after exposure to significant risk. Integrating the work of many resilience researchers, Luthar, Cicchetti, and Becker[29] discussed three main aspects of resilience that are common across multiple disciplines: (1) experiencing success despite exposure to risk; (2) dynamic adaptation to risk in order to maintain competence in adverse conditions; and (3) having a positive adjustment to trauma or other negative experiences.

Early studies of resilience focused on examining individual resilience, that is, personal traits associated with resilience. As research extended beyond at-risk individuals to multiple adverse conditions such as poverty, community violence, contagious diseases, and disaster events, resilience came to be viewed in terms of the interplay of risk and building capacity processes over time, involving individual, community, and larger social and cultural influences.[30] In Buckle, Mars, and Smale's view, "There is still a limited understanding of what the [term] . . . resilience include[s]. One of the major challenges inhibiting agreement upon any definition is due to the fact that individuals, groups and communities may each possess [differing] degrees of resilience which will vary over time."[31] Resilience is described as a dynamic process of positive adaptation in the face of significant adversity or trauma that establishes the interrelation arising from the community at risk and the wider environment.[32] Such conceptualization of resilience and its implementation provides profiles of dynamic community-environment interactions reflecting adaptive responses to disaster.[33] Positive adaptation is usually defined in terms of observable competence, or success, in coping with adversity or trauma as quickly and effectively as possible.[34]

With respect to this shift in emphasis, the concept of resilience provides a more detailed knowledge of the factors and processes enhancing communities' ability to cope and overcome adversity or disaster events. With respect to appropriate policy interventions, the resilience framework underlies emphasis not on negative consequences but on areas of strength, such as community education, community communication and warning dissemination, and community funding and planning.[35] Policy instruments that address community resilience targeted to harness notable strengths of "vulnerable communities" derive significant momentum for positive and optimistic change, such as communicative, educational, and inclusive coordination with nongovernmental entities, namely, NGOs, local leaders, and community-based local service.[36]

Identifying and building on the strengths of at-risk communities can promote their self-realization of competence and capabilities and can stimulate enduring positive changes.[37] An example of the application of resilience strategy in disaster management can be viewed from Honduras's experience of ten hurricanes and storms during the period from 1969 to 2001. Prior to the destructive hurricane of 1998, Honduras developed self-help training programs and redevelopment planning in risk management involving at-risk communities and families. Communities were involved in finding evacuation and relocation solutions such as schools and other protected buildings.[38]

Despite the significant advantages raised here, the resilience strategy suffers from drawbacks similar to those found in resistance and sustainability. In particular, resilience is more tied to natural rather than civil, economic, or political upheavals. Second, resilience strategies seem to be limited to corresponding vulnerabilities in the short run instead of reducing future vulnerabilities, due to lack of integration with central bureaucratic agencies.[39] Policy interventions should aim to address interlinked problems (social as well as technical), plan for integrated administrative and community service support at all management stages, and create opportunities for successful coordination to produce beneficial long-term outcomes.

Summing Up

This chapter identified the strengths and weaknesses of different vulnerability-based doctrines in disaster management. In particular, we evaluated how resistance, sustainability, and resilience doctrines meet the vulnerability criteria set down by McEntire et al.[40] All the doctrines share attentiveness to the role of community coping capacity in disaster management. However, the important differences between the doctrines lie in the role of public sector, integration of normative evaluation, and the interaction between different actors engaged in disaster management. Table 3.1 presents an assessment summary of each disaster management doctrine. In evaluating the disaster management doctrines, the resistance doctrine failed to integrate norms and shared values used by citizens to hold the public sector accountable, and the type of interaction existing between different actors involved in resistance management. The sustainability doctrine considers the notion of norms and social values, but it does so without referring to the role of the public sector and the relationships between different actors. The resilience doctrine seems to meet most of the vulnerability criteria by more clearly capturing the normative principles and the interaction between actors that affect the reduction of vulnerability, but is less attentive to the role of the public sector.

The comparison suggests that disaster management doctrines address the crucial role of communities and the need to incorporate community attitudes in evaluating policy options. However, this may be too crude to take adequate account of the role of community in disaster management decision-making. Considerations of fairness and equity play an important role in how citizens perceive service delivery. For instance, recent evidence from various disaster events shows that some of the communities who ascribe blame for the failed response to the government agencies actually suffered from social and racial vulnerabilities.[41] It is then suggested that examining disaster management doctrines in terms of social justice is valid and adds a new dimension to the normative assessment of disaster vulnerability.

This book contributes to the debate on social justice issues in disaster management and for policy makers provides a doctrine that better meets vulnerability criteria.

Thus, this book proposes that a resilience doctrine can bring a new impetus to the development of disaster management policies aiming to promote the well-being of effected communities by adding the social justice dimension to the normative evaluation of administration and society interactions in the face of adversity. It is argued that community trust and collaboration with public administrators cannot be maintained

Table 3.1: Comparison of Disaster Management Doctrines

Doctrine/vulnerability characteristic	Role of the Public Sector	Role of Community	Dynamic and Interactive process	Integration of normative evaluation
Resistance	√	√		
Sustainability		√		√
Resilience		√	√	√

if public administrations are unable to respond to disaster in an effective and equitable way. The respect for people is extended to respect for their experience and local knowledge, and their abilities and capacities as community members to contribute to decision making and planning at the community level. It is claimed, then, that public officials working in and with communities must be truthful and transparent in the work they undertake and the relationships they establish. Building and maintaining trust in relationships with communities should result from administrative integrity, that is, the duty to tell the truth and be honest, and engaging in surveys and delivering the findings back to community leaders, instead of approaching disaster vulnerabilities with overwhelming uniformity, utilizing universal concepts and procedures.

Thus, to maintain consistency with the normative dimensions of vulnerability and to promote effective interaction between the public officials and at-risk communities, this book offers to use communitarian ethics, which seem morally instrumental in creating an environment in which those in need feel safe enough to take responsibility for their own well-being and to assist practitioners in recognizing and identifying alternative choices when implementing comprehensive vulnerability management policy.

The revised resilience doctrine for disaster management presented in this book implies quite a change from the ordinary mode of disaster management—it pays more attention to the social justice issue of how government and administrations strengthen their communities, which means very basic capacity building, including strengthening and supporting groups and individuals involved in building the bonds of community. For this purpose, I offer to apply communitarian social justice in both public administration ethics and resilience disaster management.

Applying communitarian social justice disaster resilience doctrine addresses the need to build a community's capacities effectively and equitably to deal with the disaster event by focusing on communitarian ethics emphasizing the communities' ties, shared understanding, and common interests to gain control over their environment as the criteria guiding the decisions and actions of public administrators.

In sum, to improve the performance of resilience disaster management practice, public administration needs to play a proactive role in helping communities articulate and meet their shared interests. For that, in the succeeding chapters we need to apply communitarian social justice principles in the professional ethics and practice of public administration by establishing consistency with the common views of public service and its role in promoting the public good. Without an ethical foundation, neither individual actors nor agencies perform their proper function. In this sense, communitarian ethics is morally instrumental in creating an environment in which those in need feel safe enough to take responsibility for their own well-being and to assist practitioners in recognizing and identifying alternative choices when implementing resilience disaster management policy. Such a task leads us to examine how community-based disaster risk management shapes public administrators' professional ethics and how it allows us to analyze the dynamic interaction between state and civil society in the process of civil socialization. Then we will be able to design a resilience doctrine in ways that are fully consistent with communitarian ethics of public administration. As follows from such a resilience doctrine, the respect for persons is extended to respect for their experience as local knowledge, and their abilities and capacities as community members to contribute to decision making and planning at the community level. It is claimed, then, that public officials working in and with communities must be truthful and transparent in the work they undertake and the relationships they establish. Building and maintaining trust in relationships with communities should result from administrative integrity, that is, the duty to tell the truth and be honest, and engaging and delivering survey findings back to community leaders instead of approaching disaster vulnerabilities with overwhelming uniformity, utilizing universal concepts and procedures.

Thus, in the next chapter we draw on communitarian social justice underpinning public administration ethics and practice. By using a social justice device to achieve effective and equitable goals, disaster management policies should take into account considerations of equity, shared public interest, and responsibility.

4
Applying Communitarian Social Justice in Public Administration Ethics

After validating the need for social justice, a dimension that should and could be applied to disaster management issues, it is necessary to raise the profile of core values and ethics of public administration as key formative concerns in resilience disaster management. Our approach requires "community" input and consideration of how it influences public administration ethics and practice. It is the communitarian ethical principle of "identifying valued forms of community and to devise policies designed to protect and promote them,"[1] which makes it appropriate to apply to public administration in handling natural disasters. In other words, an applied communitarian ethics approach in disaster resilience is both normative and a political framework for promoting administrative ethical engagement because of its emphasis on social bonding and participation and, crucially, its commitment to community values and social meanings of needed goods to guide policies designed to protect and promote them. These ideas appear to resonate in the practice of disaster resilience.

Because the communitarian ethic directs the individual to take into account the source of his or her identity, and directs institutions and organizations to take into account the sources of their identities, the introduction of ethical doctrines governing public administration professional ethics is pertinent to the present study. This chapter suggests that the partiality ethical doctrine has greater potential to establish the social and professional responsibility and commitment needed between public administration and civil society than the impartial ethical doctrine that traditionally governs public administration ethics.

Professional Ethics in Public Administration

It is in small individual acts expressed through a set of relationships that the public service ethos comes to light. The manager gives expression to the ethos through dealing with people in terms of care, diligence, courtesy and integrity. The public service ethos is best perceived through the quality of these face-to-face relationships, through processes as much as results.[2]

In attempting to outline a conceptual foundation for the public administration profession it is useful to first answer the question What is public administration?

In conceptualizing public administration as a profession, I will address the issue of professional ethics both in general and as it applies to public administration. As a somewhat idealized analytic framework, I propose Asa Kasher's definition of the purpose of each profession—to promote a particular, valued goal of great importance to people's well-being in terms of principles and rules to guide proper behavior that he labels a "practical idea of professional activity."[3] While doctors properly aim at the health of their patients; lawyers, at legal justice for their clients; and teachers, at the education of their students; we say that each of them performs professional acts within professional practices. Although professional acts differ from each other in many respects, we use the label of professional for all of them. We arrive at the notion of professional practice or actions by identifying professions, not with specific groups of people but rather with certain professional practices. In other words, teaching is not what teachers do; teachers are those people who participate in the professional practice of teaching. Each professional act is performed in a distinguished context of action. Thus, in order to function in any professional area, each profession requires the knowledge and skills necessary for working toward the relevant value. For example, physicians must learn anatomy and physiology and to prescribe medication, because these skills are necessary to the pursuit of health; and lawyers must learn to set up legal documents and muster evidence and arguments to present in court, because these are necessary actions to achieve legal justice for a client. The possession of proper knowledge and special skills brings the ability to solve ordinary and extraordinary problems under certain circumstances of professional action. However, knowledge and skills alone are insufficient; within the context of action, skills alone do not possess any overarching value commitment. Indeed, it would be confusing to try to define professions by skills because the same concrete skills may be used by different professions in pursuit of their own organizing values. What counts in terms of the definition of a profession is not so much what the skills are but what they are being used to do. For example, physicians are expected to have a concept of their area of professional activity, which is to be able to understand the essence of medicine as a vocation. The extent, structure, and success or failure of medical interventions in the medical profession can be understood only in terms of the knowledge, skills, and underlying global understanding of healing that properly regulates that profession's activities. This global understanding leads to viewing the professional practice in the full honorific sense, in terms of the basic values of that professional practice, and then to regulate such activity in terms of guiding principles and rules of behavior.

The understanding of the essence of professional practice is the major aspect in identifying one profession as different from all others. For example, the medical and nursing professions are different, yet they share the same value of healing. They must be differentiated on the grounds of specific rules and principles in the pursuit of the patient's health. These principles and rules should spell out the nature of the professional practice of a given profession. Generally speaking, we need to put the notion of professional practice under philosophical scrutiny.

In discussions of the conceptual foundation of public administration, the various features of public administration as a profession, such as the profession's essence, function, purpose, mission, or goal, have been formulated. However, to the degree that the organizing value of public administration can be made more explicit than it has been, the profession can have a clearer target at which to aim and a better chance to realize its goals.

Impartiality and Partiality Doctrines of Public Administration Ethics

The study and practice of public administration has been long dominated by the impartiality approach of a neutral role for public employees in a democratic society.[4] This approach underlies the belief that bureaucrats should carry out policy directives but not influence policy creation despite their crucial role in the policy process.[5]

Advocacy on behalf of impartiality in public administration ethics, which builds on claimed synergies between public officials' goals and practice within the public sector, has made greater inroads into mainstream public administration ethics than advocacy, which argues for these goals and practices on partiality grounds. The fact that in the modern state there has been an enormous increase in the scope of governmental activity and in the range of its objectives, led to assigning immense power and authority in the hands of public officials, thereby increasing opportunities for abuse of power and authority, as well as incidents of unethical activities. Adopting an impartial stance in public administration ethical decision-making requires employing a calculus based on viewing all agents as equal and "faceless," while the particular identities, circumstances, and partialities of agents are secondary or irrelevant.

In contrast, partiality in ethical decision-making is grounded on relations arising in the context of an agent's particular, personal point of view. The application of partiality to public administration offers normative views of the role of the public administrator rather than views that are instrumental and value-free. It is claimed that the characteristics of public service are such that they raise ethical issues that are somewhat unique: public administration is a profession combined with work in the public interest; the officeholder must adhere to higher standards of conduct than others in society.[6] Indeed, public officials often experience conflicting ethical decision-making, such as whether to give precedence to the public interest or to the narrower demands of profession, department, bureau, and personal interests.[7]

Within the public administration literature, partiality toward others has become devalued, and indeed actively discouraged, because growing inappropriate use of partiality by people who hold public office in ethical decision-making, such as nepotism in governmental office (bestowal of patronage positions by reason of relationships rather than merit), misappropriation (illegal allocation of public resources for private use), and official corruption, has provoked great consternation.[8] These phenomena express concern that something is wrong with the unqualified practice of partiality, leading to less integrity and fulfillment than is desirable.

Impartiality Doctrine

Impartiality concerns the nature of moral deliberation aimed at removing the biasing influences of one's objectives, interests, and favoritism based on the agent's personal characteristics, background, values, and beliefs.[9] In order to make sound moral judgments, the demands of impartiality require invoking independent points of view. Thus, the significance of moral impartiality is seen as arising from the fact that a core role is given to its universal applicability. An appealing justification of this view of morality was given by deontological (Kantian) ethics and utilitarian ethics.[10] In both deontological and utilitarian ethical theories, impartiality finds its ultimate grounding in rights, duties, and utility to society. The process of moral reasoning based on rights and duties is designated as deontology (from the Greek *deon* = obligation). Deontological theories hold that actions are based on duties or principles that are intrinsically right or wrong and thus obligatory or forbidden, regardless of the motives for which they are performed or the states of affairs in which they result.[11] Following the utilitarian form of teleological ethics, acts are obligatory if they meet the test of the principle of utility; that is, if they maximize the happiness of a larger number of people than would alternative courses of action.

Studies in public administration ethics have attempted to produce hybrid systems in which deontological and utilitarian theories are brought together. George Frederickson, Mary Guy, and Carol Lewis[12] suggest reconstructing teleology and deontology not as mutually exclusive models, but rather to entail a determination of ethical behavior in which both the consequences and the intrinsic character of an act may be included in interrelations. These scholars paved the way to consider a more unified ethical decision-making model for public administration, including Ralph Chandler and Gerald Pops,[13] who have tried to account for deontological forms of teleology by calling them "mixed" teleology and deontology rather than distinct models. Garfalo and Geuras proposed viewing deontology and teleology as different ways of seeing the same thing; for example, the need to promote social equality may reflect a commitment to the principle of justice and improving the living conditions of a minority group;[14] and Svara proposed the ethics triangle, combining more dimensions in ethical decision-making in public administration, namely duties, goods, and virtues, in order to reduce the complexity of ethical reasoning by formulating general principles.[15] Indeed, in most integrated ethical models, ethical principles or judgments are based on impartial deliberation. From a lexical ethical view, impartiality prevails over other doctrines and serves as the dominant doctrine to guide ethical thinking and conduct in public administration.

Within the discussion on the application of impartiality in public administration ethics and practice, one can benefit from the comparison between impartiality and neutrality. At first sight it seems that "impartial" and 'neutral" are synonymous and thus share some definitional common ground. While the absence of assistance and attachment implicit in each term is very similar, there is an important difference in their respective detachment from all parties. A neutral stance is far more passive than an impartial one. Neutrality is an "absence of decided views," while impartiality is "fair and just." Neutrality is a passive "vague" and "indefinite" position, whereas impartiality is a coherent position capable of demonstrating an absence of prejudice, favoritism, and bias.[16] The danger to impartiality arises when we take for granted as starting points for analysis that which we already know well, without pausing to justify it.[17] In other words, when we identify a problem, we inevitably use our categories to tell us what to do. Thus, the comparison between impartiality and neutrality sheds light on how impartiality can reveal a mindless repetition of the nonneutral past. Impartiality, then, requires us to look for starting points and expose them to questions, to recognize the partiality of every judgment. It also means securing against disqualifying a judgment merely because it is partial. It means taking into consideration the partiality of everyone's perspective through consultation and collaboration, which is halfway toward treatment of a policy problem.

The Partiality Doctrine

Partiality, as an ethical reason, pertains by virtue of a relation between an agent and a particular object of value.[18] Ethical judgment arises from the personal point of view of the agent, shaped by the background context of justice of an agent's particular, personal, point of view. The demands of partiality thus stress the moral value of the varied sorts of personal relationships in which it is featured.

The beginnings of partiality in ethics go back to the late 1950s, where scholars criticized the dominance of impartiality as the master principle in moral deliberation. Most of them, like Philippa Foot and Elizabeth Anscombe,[19] belonged to the camp of (neo-)Aristotelians. They claimed that the nature of moral deliberation is constitutive of agents' identity, and therefore cannot give precedence to the requirements of impartiality. Partiality is the process of reasoning with others as the core of ethical decision-making, invoking many aspects of partiality, including relationships[20] in general, the unique nature of circumstances and the parties involved,[21] self-knowledge,[22] political community and tradition,[23] and care-giving relationships.[24]

During the last decade, scholars have explored the adequacy of normative aspects such as equity, justice, and fairness in explaining administrative decision-making.[25] These studies aim at showing how contested ethical issues arise in those situations where public administrators are "acting at a distance" in order to implement policy for those in need within one's own society, in situations they may not understand.[26] By situations, public administration ethicists mean that situations and social conventions assign responsibilities for the care of those who need it to particular others as an attempt to ensure that everyone is cared for at dependent stages of their lives and that, in emergencies, there are guidelines that ensure effective help to those in danger.[27] Thus, public administrators are often uniquely situated to answer someone's need, which derives from an ongoing relationship with that person by virtue of which one has been held responsible for his or her well-being. In this sense, partiality is morally instrumental in a relationship to the extent that it contributes to the protection of those who are in need and on social conventions that assign responsibilities for the care of needy persons to others who stand in certain relationships to them.

Within the discussion of when it is appropriate for public administrators to act on partiality, and when on impartiality, scholars endorse the view that none of these doctrines gets priority when a partial principle conflicts with an impartial principle.[28] Recognizing the ongoing discussion of the applicability of impartiality or partiality as a justified doctrine in public administration ethical deliberation, this chapter offers some progress in the debate concerning the context of social justice from which we can better evaluate them.

In evaluating these ethical doctrines this chapter holds the assumption that partiality has the potential to produce a more equitable distribution and access to public good than the impartial doctrine that traditionally governs public administration ethics. In my view, commitment to universal value such as public interest shapes public administration ethics without which public administration is merely an instrument of its political master. Public administration by its nature is ethically bound to the society, which expresses its values by entrusting them to the public sector. Therefore, defense of the administrative practices by which such partiality would be normalized and turned into a set of legitimated expectations for a society will be derived from a sense of integrity and fulfillment that is attained in close relationships, to the extent that personal relationships are necessary for integrity and fulfillment in life.

Social Justice Foundation for Public Administration

As seen, we consider public administrators involved with disaster relief efforts to be professional "helpers," as their aim is not the allocation of goods to the deprived, those who fall below the minimum just level of basic goods, but the enhancement of recipients' and potential recipients' lives by giving them a greater say in which goods should be distributed.[29] Public administrators should help citizens articulate and meet shared interests. This is not only a matter of institutional constraint on public officials in disaster management, but also shapes the distinctive nature of the public service profession as a whole. Authors in this area have proposed several models of the citizen-administrator relationship based on different views of the nature of society and the place of government and public service in our lives.[30] Despite the fact that these scholars would resist characterization as communitarians, aspects of their work are similar to the communitarian vision of administrators in facilitating citizen activism or even taking an activist role in local communities.[31]

From the social justice viewpoint, the goal of administrators as professional helpers is to provide needed resources that promote the abilities of communities and individuals to pursue their own distinctive perceptions of their needs. Perceptions of basic rights and goods surrounding disaster management policies are often at the core of social justice disputes. Local officials should encourage affected communities to participate in formal decision-making processes throughout the course of preparing action plans led by higher-rank officials, so that affected communities will be able to greatly reduce the total amount of damage and vulnerability caused by disasters.

The task undertaken here of clarifying the complex relations among social justice and public administration is critical for moving the field toward a unifying conceptual foundation. It is argued that disasters are powerful indicators of the state's relationship with the most vulnerable members of society. Thus, communitarianism gives an account of the nature of justice as we already intuitively understand it, or, failing that, as we could be brought to understand it if we took a rational and sustained interest in doing so.

Applied Communitarian Ethics in Public Administration

Communitarianism begins with one of the most important moral insights of ancient through modern times and couples it with a powerful metaphor that underlies our moral life. The insight is that engagement in society makes us what we are. The idea of community having intrinsic value was addressed in Aristotelian thought, and identified later in the contributions of Rousseau, Montesquieu, Burke, and Durkheim. Aristotle viewed the "polis" as an end of human association, in which human relationships find their fulfillment. Thus, the community provides individuals opportunities to establish satisfying filial relationships, become educated, practice a profession, discover one's spiritual aspirations, join a fraternal association, and, most importantly for Aristotle, establish close personal friendships. For Aristotle, he who lives outside the community is "either too bad or too good, either subhuman or superhuman."[32]

Viewing a being without others as nothing more than an animal, puts this insight in sharp contrast to the classical liberalism in the pursuit of individualism.[33] While classical liberalism holds no bound between people in any kind of collaborative arrangement, communitarianism is concerned with people bound together in some kind of collaborative arrangement, being of the same nationality, subject to the same laws, governance, and so on.[34] For instance, Walzer, Sandel, and MacIntyre hold that justice is at least partly relative to the particular culture and history of each society.[35]

Quite often communitarianism is presented strategically as a rival approach to liberalism, rather than as an internal criticism of liberalism.[36] However, it is suggested that both modes of social justice fall explicitly within the social contract tradition, but diverge from each other by legitimizing different bases of fairness for social cooperation in the context of an unavoidable pluralism regarding views about the good life.[37] For instance, Taylor[38] suggests that this apparent polarization between what might be called the universalists such as Rawls and Dworkin, and the contextualists[39] such as MacIntyre,[40] Walzer, Sandel,[41] and himself, is the result of misunderstanding that a disengaged view is either possible or necessary for determining the justice of particular distributions. Taylor claims that what goods are distributed and to whom will depend on particular associations, which create the context within which they are achieved, "a hyper Kantian agent capable of living by rules which utterly leave out of account his or her advantage and which could be agreed by everyone since they are not designed for anyone's good in particular."[42] Rather, the priority in the way that the distributive principles are applied is contextually specific.[43]

Michael Walzer defended the communitarian view by adding that morality and our sense of self cannot be captured in a vacuum. Walzer stresses a deeper sense of what society—and us along with it—can be, and wants to lift us to our true potential greatness in comparison to the liberal sense of society.[44] His aversion to the moral universalism presented by liberalism lies in his substantive disagreement over the disaggregation of society to isolated selves:

> The world is not like that, nor could it be. Men and women cut loose from all social ties, literally unencumbered, each one the one and only inventor of his or her own life, with no criteria, no common standards, to guide the invention: these are mythical figures.[45]

Walzer then sets the link between an individual's moral choices and the legitimacy of coercive state intervention through the understanding of the individual within the context of community—community being understood as family, friends, the tribe, the neighborhood—of people with whom an individual personally interacts.[46] As such, community is conceived as a mediator construct that provides the conditions and the background for distribution of goods. Goods obtain meaning in terms of the community value they have. Distribution and exchange create meaning materially and symbolically: "The Idea of distributive justice presupposes a bounded world within which distributions takes place: a group of people committed to dividing, exchanging, and sharing social goods, first of all among themselves.... The primary good that we distribute to one another is membership in some human community."[47] Thus, goods obtain meaning because they are shared by community members, because they have a particular function in a given society: "By virtue of what characteristics are we one another's equals? One characteristic above all is central to my argument. We are (all of us) culture-producing creatures; we make and inhabit meaningful worlds."[48]

The central assumption laid by communitarianism concerns the "Social Constitution" that refers to viewing beings as persons derived from the existence of community. This means that personhood is essentially dependent on community, and that we can lose our nature as persons. Communitarianism then goes beyond the biological roots of the sociality of human beings by referring to nature as serving only a predisposition to community.

The understanding of social constitution is offered by MacIntyre to denote the social ties that "constitute the given of my life."[49] Taylor claims that the individual possesses his or her identity through active participation in community,[50] and Sandel insists on identifying such ties as constitutive "attachments" that we do not "voluntarily incur."[51] Sandel maintains that being born into and developing within a community with others entails individuals with constitutive ties that shape our identity: "[C]ommunity describes not just what they have as fellow citizens, but also what they are, not a relationship they choose ... but an attachment they discover, not merely an attribute but a constituent of their identity."[52]

At the heart of the communitarian ethic is the relationship between the self and its community. The communitarian concept of the relationship between self and community is conceived differently from the liberal-individualist one, as stressed by Kymlicka:

> The community has no moral existence or claims of its own. It is not that community is unimportant to the liberal, but simply that it is important for what it contributes to the lives of individuals, and so cannot ultimately conflict with the claims of individuals. Individual and collective rights cannot compete for the same moral space, in liberal theory, since the value of the collective derives from its contribution to the value of individual lives.[53]

The communitarian self is articulated in Sandel's work as

> [T]he image of the unencumbered self is flawed. It cannot make sense of our moral experience, because it cannot account for certain moral and political obligations that we commonly recognize, even prize. These include obligations of solidarity, religious duties, and other moral ties that may claim us for reasons unrelated to choice.[54]

He stresses once again the idea that the "vision of the person . . . inspires and undoes this ethic": "the predicament of liberal democracy in contemporary America may be traced to a deficiency in the voluntarist self-image that underlies it."[55] There are two arguments underling Sandel's theory. The first argument rests on the idea that liberalism (grounded in the deontological ethic, which is concerned with rules and obligations) neglects to appropriately conceive persons and their (our) moral experience. The second argument concerns the "empowering" attachments necessary for flourishing individuals who are

> free to choose our purposes and ends unbound . . . So long as they are not unjust, our conceptions of the good carry weight, whatever they are, simply in virtue of our having chosen them . . . This is an exhilarating promise, and the liberalism that it animates is perhaps the fullest expression of the enlightenment's quest for the self-defining subject. But is it true? Can we make sense of our moral and political life by the light of the self-image it requires? I do not think we can.[56]

Sandel's reflection on the relationship between the self and its community draws on two seemingly opposing conditions: the social constitution of individuals as well as their independence from the community. These positions are both articulated in the communitarian's commitment to the social constitution thesis that incorporates individuals' autonomy to the extent that so long as they participate in community, they can exercise their autonomy and self-determination.[57] Both conditions generate a "deep self" consistent with the social constitution thesis that is independent of community, even though the "whole" or "greater" self is not.

As a theory of people in relation with each other, communitarianism asserts that society exists prior to the individual and that it creates the social self. The assumption that persons are socially constituted provides continuity of the life-world, allowing individuals a place and time within which to function and exercise their capacities through interaction with others, resulting in interdependence. From this interdependence flow the "primordial sources of obligation and responsibility."[58] The social embedded self consists of both an "I" and a "we" through the interactions with others that go against the liberal image of the unencumbered self:

> In deciding how to lead our lives we all approach our own circumstances as bearers of a particular social identity... Hence what is good for me has to be the good for one who inhabits these roles. Self-determination, therefore, is exercised within these social roles, rather than by standing outside of them.[59]

Understood in this way, communitarianism implies that society or communities contribute to a person's constitution through active participation of its members. Greater emphasis should be paid to those constitutive ties that hold individuals together. Acknowledging that community constitutes one's sense of identity and relatedness entails certain obligations to the community and its members. Therefore, strong communities whose citizens feel moral obligations and a sense of solidarity with each other can only exist when the members share a common form of life. This is only possible when the idea of the neutral state is replaced by a nonneutral "politics of the common good."[60] The community defines the common good based on a "feeling of commitment to a common public philosophy... as a precondition to a free culture."[61] In this sense, communitarian ethics prescribes that the responsibilities that are anchored in community grow out of diverse moral sources of citizens in order to create free and open dialogue between community members.[62]

As stated by Bellah:

> A good community is one in which there is argument, even conflict, about the meaning of the shared values and goals, and certainly about how they will be actualized in everyday life. Community is not about silent consensus; it is a form of intelligent, reflective life, in which there is indeed consensus, but where the consensus can be challenged and changed—often gradually, sometimes radically—over time.[63]

The public interest results from a dialogue, concern for the larger community, and commitment to assume responsibility for their community rather than mere aggregation of individual self-interests.

Communities, then, share common meanings and values within their language and actions. The legitimization of a community's values rests not on consent but on what sociologists call the implicated self, an idea that postulates that "our deepest and most important obligations flow from identity and relatedness, rather than from consent."[64] Surely, relatedness entails duties to others; within this context the duty to respect the rights of others arises.[65] Thus, unlike liberalism, which holds the values of autonomy and individual rights with few social restrictions, the thin social order communitarianism requires a precondition to freedom and rights that refers to a society that possesses common values to justify many reasonable restrictions on the individual to protect those values: the thick social order. In other words, community members have a duty to defend their shared values when under attack by others from within, because failing to do so would cause the "debasement and decay" of the community's values and ultimately the community itself.[66] The objective of communitarian society is articulated by Beiner as

> The central purpose of a society, understood as a moral community, is not the maximization of autonomy, or protection of the broadest scope for the design of self elected plans of life, but the cultivation of virtue, interpreted as excellences, moral and intellectual.[67]

To conclude, communitarianism concerns individuals living in a community where they exercise their free will but where personage is shaped through a common language, values, and concepts that in turn provide them with sense of responsibilities, ethical conduct, and civic virtue. The communitarian vision of civic virtue and responsibility calls on institutions of civil society to occupy a central role in reflecting the critical and natural expression of social men and women in relation to each other and their mutual responsibilities.[68] For that, communitarians believe that the intimate relationships between human beings and all their institutions generate social capital; these critical connections are essential to nurture obliging citizens with moral and virtuous purposes.

The conceptualization of communitarianism and its application in ethical thinking is not without criticism, and major weakness is evident on the part of the integration of both independent and autonomous aspects of individuals and the enhancement of the common good generated from engagement in community. In particular, concerns have been identified regarding the subjective and often unarticulated claims underlying the objective to strengthen community bonds. The adherence to communal values may, in various contexts, justify the promotion of cohesion and authoritative means imposed on individuals to protect those values.

Disaster as a Catalyst of the Community and Public Administration Bond

This section focuses on the ways public administration may address community and civil society in an environment of disasters. It raises the question of whether public administration should work to encourage citizen identification with community or the civil society at large. I want to turn that focus, and the related question, so that civil society, or more specifically community, can provide public administrators a means of dealing with the wide range of challenges posed by the disaster events.

In times of adversity, the social bond and solidarity is increasingly under stress, but its function is critical for any sort of effective collective action to take place. At the same time, the strength of performance of public administration can contribute to building civil society capacities.

The term *civil society* has gained numerous definitions since the 1980s. What is common to most of the definitions is the distinguishing nature of the civil society from the state consistent with the understanding of civil society as located between institutions of the state and those of the private sector. For example, Diamond, Linz, and Lipset[69] offer an inclusive definition of civil society as

> the realm of organized social life that is voluntary, self-generating, (largely) self-supporting, autonomous from the state, and bound by a legal order or set of shared rules. It consists of a vast array of organisations, both formal and informal: interest groups, cultural and religious organizations, civic and developmental associations, issue-oriented movements, the mass media, research and educational institutions, and similar organizations.[70]

Under such definition, economic, religious, kinship, and political interest groups are all considered as part of the civil society. A more exclusive definition of civil society is provided by Stepan as,

> that arena where manifold social movements (such as neighborhood associations, women's groups, religious groupings, and intellectual currents) and civic organisations from all classes (such as lawyers, journalists, trade unions, and entrepreneurs) attempt to constitute themselves in an ensemble of arrangements so that they can express themselves and advance their interests.[71]

Following this definition, political organizations are not included in civil society but rather create a separate space called "political society." Another exclusive definition of civil society is suggested by Huber, Rueschmeyer, and Stephens,[72] as "the public sphere distinguished from the state, the economy, and the web of family and kin relations."[73] Economic organizations are not included under this definition, although political organizations are.

Schmitter offers to view civil society as a more intermediary organization located between the state and primary social units of production and reproduction due to its autonomy and capacity for collective action.[74]

Other definitions of civil society underlie values rather than space. Civil society organizations share norms and values to encourage and enable individuals and groups to take initiatives and to carry through independent actions. Given the limited numbers who participate in the political system through voting opportunities, let alone engage in participatory or civic actions, civil society holds values and norms of solidarity, advocacy, collectivity, trust, mutual help, tolerance, and so on, facilitating all varieties of civic participation in political life at the community level.[75] Putnam acknowledged the values of civic engagement that enable individuals and groups to build communities, to commit themselves to each other, and to bind the social fabric.

For the purposes of this book, we focus on civil society in terms of both space and values as a sphere of association and action independent of the state and the market in which citizens can organize to pursue social values and public goals that are important to them, both individually and collectively.[76] However, the conceptual separation of civil society from the state does not mean pitting civil society "against the state." The interaction between the state and civil society is justified for civil society to function effectively and to alleviate conflicts with the state. Several scholars acknowledged that rather than a zero-sum game, the establishment of civil society and state relations would lead to a win-win situation, when "Civil society and the state must become the condition of each other's democratization . . . [including ultimately] attempts to democratically expand civil society through state support."[77] A positive-sum relationship is regarded as a critical ingredient of participatory democracy. This form of relationship creates favorable conditions necessary for citizens to regularly engage in political life— and not only during elections: it creates a framework for citizens to articulate their interests and make demands, thus contributing to the development of a vibrant democratic society; it forces state agencies to become more transparent and closer to their constituencies, and it facilitates the effectiveness and the implementation of public policies. This potential of a win-win situation also leads to widening the scope of civil society-state relations and interactions, as emphasized by Young:

> Though civil society stands in tension with state institutions [conceptually, in practice] a strengthening of both is necessary to deepen democracy and undermine injustice . . . social movements seeking greater justice and well-being should work on both these fronts and aim to multiply the links between civil society and states.[78]

Such an understanding of the relations between civil society and state is apparent in the public administration literature. Focusing on the relations between civil society and public administration touches on the very question of to what extent bureaucrats' actions should meet the demands and needs of civil associations or the citizens at large. In the 1940s, Friedrich maintained that professional bureaucrats should hold responsibility by encouraging accountability and an embedded sense of the public interests.[79] Maas and Radway distrusted the idea that administrative agencies should have to be responsive directly to citizens at large. Instead, they offered to narrow the scope of bureaucratic accountability to pressure groups entrenched with the knowledge needed for effective policy implementation.[80] In 1987, Romzek and Dubnick examined the Challenger Disaster of 1986, in which the space shuttle exploded, as a tragedy that resulted from poor management of political and bureaucratic accountability mechanisms that hinder the technical and professional capacity of NASA to pursue its goals. Similar themes were pursued in the 1990s by Stivers, who describes the polis as "a public space in which members act together in order to achieve limited ends and to lead a virtuous life." [81] Stivers has laid the groundwork for resolving the tension between professional accountability and democratic accountability experienced by civil servants in developing their listening skills. The "Listening Bureaucrat" can be open to citizen and group voices, communicate with stakeholders, create a shared public or common space, and accept different voices and perspectives. According to Stivers, the practice of listening promotes accountability as it develops openness, inclusion, respect for differences, and mutual commitment that supports both democratic accountability and administrative effectiveness.

Frederickson said, in part, that administrative agencies should be established on principles of inclusion to be engaged with greater levels of citizen associations.[82]

To improve civic participation in the policy process, Crosby, Kelly, and Schaefer suggested the utility of citizen panels to overcome the tensions between public and professional accountability.[83] De Leon called for increasing citizen participation in the articulation and formulation of public policy as an integral part of the public sector reforms during the 1990s, known as the New Public Management.[84] In 1992, Rourke referred to the political pressures imposed by the presidents and the U.S. Congress on bureaucratic action and responsiveness. Rourke's study does complement this review by bringing attention to political constraints on civil society and administration.

Taken in total, the theoretical framework described here indicates the need to further investigate the relations between civil society and administration by addressing the significant systemic and structural barriers to effective and authentic civic governance processes in the face of disaster. Consistent with the idea of potential positive-sum relations between civil society and administration, public officials need to adopt a broader and longer-term perspective that requires a knowledge of effective means to contribute to building a collective shared notion of public interest; that is, a sense of belonging, a concern for the whole, and for vulnerable communities in particular. Following this view, citizens and communities are not merely customers or clients but support the connection between citizens and their communities that is the essence of community building and of civil society and democracy as well. In short, it is suggested that public servants will take an active role in creating the space in which citizens and communities can articulate shared values and develop collective action on behalf of the public interest. Public administration must respond to citizens as defined by their commitment and responsibility to what happens in their community rather than simply in a legalistic sense. Nowadays, more than ever, the rise of information technology opens the door to citizens' minds and web-based tool kits that can mobilize needs and solutions from the citizens to the civil service. E-democracy becomes a valuable tool for citizen integration in policy making.

In sum, I believe that a normative-oriented approach to public administration has the power to make sense of public administration as a distinguished profession in ways not yet generally appreciated. Public administration can be conceived of as a profession engaged in alleviating vulnerabilities in all its forms, from economic to psychological; public officials identify people who are at risk in any justice-related good and intervene in order to help them gain their capacities to cope in the face of adversity.

In the next chapter we take a further step in applying communitarian ethics set by Michael Walzer for practical engagement of public administration in resilience disaster management. The use of Walzer's communitarian framework to conceptualize public administration practice will proceed by showing how the value of distributive justice is part of the relatively unrecognized deep structure that generates public officials' judgments. The purpose of such analysis is to enable public officials to see more clearly what the profession already is and to work more explicitly, and thereby more effectively, toward the goals to which they already aspire.

5
Walzer's Communitarianism in the Service of Disaster Resilience Doctrine

As suggested in the previous chapters, disaster management policy directed at building and improving the resilience of communities at risk needs to ensure positive adaptational outcomes and their antecedents. From a policy perspective, this implies an emphasis on public officials' efforts to identify and build on the strength of communities in emergency circumstances, with the aim of supporting and promoting feelings of competence and capability among its members. Thus, the question at issue here remains how to encourage communitarian ethical engagement in specific administrative choices and practices, along with their resilient consequences. It is argued that attention to the communal provisions in Michael Walzer's theory of communitarian justice are particularly relevant in light of the central role community occupies in disaster programs and policies. Thus, measures of enhanced resilience, including advocacy, diversity, competence, and inclusion, are aligned with a communitarian ethics framework to help in building relationships between public administrators and vulnerable communities.

This chapter thus suggests that issues surrounding the management of disaster resilience involve a communitarian framework where needs-meeting is functional for resilience policy in natural disaster events. For that purpose, we will draw on the principles underpinning public administration resilience disaster management by applying Walzer's theory of communitarian justice, and consider how these influence disaster management objectives, including (1) advocacy, (2) inclusion, and (3) competency. These attributes emphasize deep social and community transformation rather than merely helping people adapt to risks and crisis circumstances (see fig. 5.1).

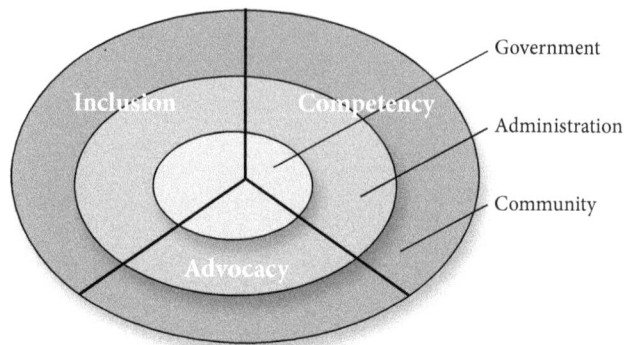

Figure 5.1: The community-based disaster resilience model

Michael Walzer's Communitarian Justice for the Analysis of Community and Public Administration Relations

Walzer's concept of justice can be extended to reflect on the distinguished role of public administration in terms of social capital. Most generally, we want to apply Walzer's core assumption—namely, that the state, for all its putative faults, is nevertheless an indispensable mechanism of the destratification of civil associations. This assumption may sound too provocative, since most recent studies in social capital theory reveal that it is the power of civil associations and groups that paves the way for democratic and equitable society.

Despite the wide variety of diverse theoretical origins and empirical applications, it is possible to define social capital as the communal inventory of "generalized trust," mutual obligations, and shared cooperative attitudes that have functional utility for individuals.[1] However, in most social capital theories, the role of public administration is devalued. Indeed, most "bridging" and "bonding" forms of social capital establish relations to others and associations *within* the same horizon or social stratum. Put simply, the bridging functions of social capital characteristically lead individuals to "bond" with and "trust" others more or less like themselves. In the same way, such "bonding" functions of social capital create *internal* bridges within the existing strata of civil society. Social capital in light of Walzer's theory is viewed as embedded in the hierarchies of civil society and thus supplies a mechanism of the (re)production of various forms of inequality:

first, because the state is necessary to enforce the norms of civility and regulate the conflicts that arise within civil society; second, because the state is necessary to remedy the inequalities produced by the associational strength of different groups . . . and third, because the state is necessary to set limits on the forms of inequality that arise within the different associations.[2]

Social capital is then featured as a highly group-specific, socioculturally and economically context-dependent form of capital circulated and deployed by associations and social groups in their everyday struggles for control over the consumption and distribution of limited economic, social, and cultural resources. Walzer's theory of distributive justice offers to capture the distinguished nature of public administration professional ethics and practice as "mediating associations."[3] What makes public administration associations "mediating," as we shall show in the next sections, is that they reflexively cross socioeconomic, cultural—and at times even ethno-racial—strata to expand community members' normative conceptions of capacity and resiliency, and at the same time they engage the larger mechanisms of the state.

It should be pointed out that there are a few limitations for the proposed framework. First, Walzer aims at constructing a highly theoretical framework with no intention to provide a basis for public administration professional ethics and practice, so that applying his ideas to public administration involves borrowing, adapting, and extending his insights. Second, all claims about the nature of justice are, of course, controversial and open to debate, so the ideas cited here should be taken as a tentative dialectical beginning and not as established dogma.

However, I believe that Walzer's theory of distributive justice can accommodate administrative practice, as it has a good chance of providing a more unifying conceptual framework for public administration deliberation in disaster resilience than now exists, which also bears implications for the process of strengthening the civil society.

The Role of Public Official as Professional Helper

Walzer's theory of distributive justice bound to his conceptualization of community of justice can be understood as requiring that a society be closed off in some manner. Thus, the first distributional sphere that

Walzer offers is membership. Walzer uses a homogenous concept of membership, which denotes that in every sphere (e.g., security and welfare, money, office, hard work, basic education, private recognition, and social bases of self-respect) access to the distributed goods is reserved to the same category of membership. Consequently, the possession of membership determines one's share in other spheres. If individuals engage in the exchange of social goods and participate in collective decision-making as "members," their relation is based on shared values that requires to some extent that they are symmetrically placed

Membership is perceived by Walzer as the only natural locus for reflecting on the overall justice of institutions. Social isolation is required for distinctive communal meanings to arise and take hold, and for cultural consciousness and character to develop. As Walzer points out, "We do not owe the same goods to every-one [sic]."[4] The boundaries of community are the boundaries of the political community, which is defined as a collectively self-governing community of political equals. So it is within political communities that individuals can best be placed to work out their assessment and articulate criticism of dominance and inequality rather than in their more specific roles of clients, professionals, and so on. Thus, the underlying assumption is that in settling conflicting issues of distributive justice, citizenship is a predominant consideration. Following Walzer, communal provisions are shaped by shared understandings of all the various things that are conceived, created, exchanged, and divided by the group. Thereby, political communities exist to provide for the needs of their members, as their members articulate those needs. Whatever goods a particular political community collectively provides, it is based on those goods that the political community determines that members owe one another, as well as the priority of these needs and the degree to which these needs are met. The priority of needs and the degree to which any particular need is met are a reflection of the particular culture, character, and understandings of the community, instantiated through political processes. Within the same community, not all needs will be met to the same degree, and it is unlikely that any particular need will be fully met. For example, individuals need to acquire some basic skills to be able to participate in communal life; they should be educated up to some minimal level of basic competence.

Being a member of a community refers to a way of relating to other people. Therefore, the boundaries of the political community entail obligations we have to other members. Obligations can arise from belonging, identity, and relatedness. For instance, the concept of reciprocity reminds us that any member of a community enjoys various social benefits for which that person is not required to pay directly. Hence, a member of a community also should bear certain obligations toward his or her community. Thus, relations of reciprocity and mutual trust, for example, generate the necessary solidarity, willingness to help others, and so act responsibly.

This social vision underlies the significance of belonging to a community of justice—special stronger obligations to others formed by shared history or close personal relationships, and perceives our membership therein in purely exclusive terms, and insists that our mutual obligations as members of such a community should be derived either from our consent or from their being to our advantage. In contrast, reciprocity for Walzer presupposes, or at least attempts to inspire, an inclusive social vision. Such a social vision maintains that obligations do not diminish in strength when we move beyond the boundary of the political community and proceed to those more geographically or culturally distant from us ("society of equals"). The appropriateness of this inclusive view of community is actualized in the sphere of security and welfare, where goods are to be collectively provided to each needy person in proportion to his or her neediness.[5] Equality does not only require equal access to defined channels of influence. It also calls for enhancement of social and political empowerment among citizens in terms of acceptable living standards, proper level of education, and other distributed resources for ensuring that all citizens are competent in articulating their interests. Once a community devotes public funds to provide a good on the basis of the need of its members, any other basis or criterion for distribution should be viewed as a distortion of the process, a violation of the shared meanings and common understandings that are at the foundation of the human society as a whole.

Taking on Walzer's concept of community, public administrators' responsibilities are rooted in one or more of their social roles, such as member of a professional association, parent, or member of a religious organization, that are essential for carrying out public responsibility in a consistent, rational, and dependable fashion. The position of bureaucrats within a society suggests special bonding between public administrators, who are at the same time recipients, and citizens (the actual and potential recipients), which ensures that responsibility promotes the common good. Following Walzer's view of community, public administration's mediating role in society is a constitutive component of the concept of distributive justice that considers public reciprocity while maintaining some sense of intimacy.

> Personal responsibility can be described in either of these ways [particular or universal]: some such idea seems to be generated by every experience of social interaction, but it can take different forms with reference to this or that good in this place or that place, with different consequences for distributive arguments.[6]

Reciprocity in public administration ethics and practice as conceptualized here insists that citizenship should not be reducible to a market relationship, by urging administrators to adhere to the plural and ambivalent understandings of citizenship—as both a source of mutual advantage and a locus of membership and belonging.[7]

Local Capacity Building: Gaining the Trust of the Community

The notion of distributive justice, according to Walzer, lies in a community of justice at its foundation. Walzer's broad construct of the scope of distributive justice is not arbitrary. It results from an analysis of community as a distributional enterprise: "The community is itself a good—conceivably the most important good—that gets distributed."[8] For Walzer, our common intercourse with other individuals naturally forms bonds and engages in social compacts to provide mutual support and to create and distribute those things that we need, such as with family members, friends, the tribe, and neighbors.[9] In order to engage in collaborative efforts to mutually provide for each another's needs, justice requires that goods will be distributed following a shared understanding of needs.

To illustrate that, a good can be said to be needed if its possession has a particular degree of importance or urgency. The particular degree or kind of urgency depends on the role the good is understood to play in their lives. The history, language, religion, politics, and so on of the community will help shape, and shape in different ways, what a particular group will conceive as necessary parts of common life. Our understanding of bread, for example, is largely determined by history and cultural context: "Bread is the staff of life, the body of Christ, the symbol of the Shabbath, the means of hospitality."[10] One can reveal the two layers of social meaning underlying the importance or urgency. In-depth investigation into the concept of social meaning can uncover two layers of social meanings that refer both to those aspects of a good that are relevant to the way it contributes to the good life of people and those that are judged to be needs.[11] Although Walzer's appeal to social meanings fails to distinguish between these layers of social meaning, it seems plausible to assert that Walzer's theory gives precedence to community membership as the initial distributed good and those aspects that are relevant to the way it contributes to the good life of people, which is, to large extent, a common life.

But how is Walzer's theory more compelling to public administration ethics than other egalitarian theories? Walzer's theory reveals the motivation of public officials for the project of distributive justice. Talk of the role of public officials as members of a political community is often associated with the virtue of social responsibility that can be accommodated within the distributive criterion that distinguishes between regulations. It suggests that such an approach constitutes a valuable instrument for imbuing our local communities with civic responsibility. Such sharing of information between members of a community can nurture trust and caring that is constituted in intimate relationships. For Walzer, the intrinsic value of community is expressed in terms of moral development of trustworthiness: "The mere fact that human beings are vulnerable and frightened and need to be able to trust one another and so on only make[s] for morality."[12]

Thus, trustworthiness is an evolutionary process in which our initial impressions of the relative trustworthiness of others is likely to be both incomplete and in need of subsequent review as we learn more about those individual others. As Potter notes,[13] "When we want to determine whether or not to trust another with the care of some good we value, we need to know what the other's values, commitments, and loyalties are." Potter's consideration of the value of trustworthiness applies the Walzerian idea of how particular goods ought to be distributed in particular cultural contexts, by virtue of how the people come to understand the meanings of those goods. People must come to recognize that government is responsible for assuring distribution of goods that is consistent with public interests and with norms of fairness and justice in both substance and process.

To work out Walzer's understanding of trustworthiness as a justice principle internal to social good, one can translate Fukuyama's description of culturally determined differences in trust relationships to illustrate differences in the development of business relationships in various parts of the world, and citizen–public officials' relationships. Fukuyama describes the cultural determinants of the boundaries of our trust relationships when, for example, he compares the different approach of Chinese communities that consider trust to be limited to family members, with the European approach of trusting people outside the family.[14] He draws on Redding, who states:

> The key feature [of Hong Kong businesses] would appear to be that you trust your family absolutely, your friends and acquaintances to the degree that mutual dependence has been established ... With everybody else you make no assumptions about their goodwill.[15]

Fukuyama's position draws attention to the claim that people are in fact limited in their choice in trusting because of their cultural inheritance, but within the Walzerian framing of local autonomy, I think he would probably allow for a partial understanding, that people's choices nevertheless rest on the idea that people will trust those whom they assume to have good will toward them.

Attending to the principle of trustworthiness in a partial way will consequently enable the citizens to gain a proper understanding of what it means to participate as an equal member in the huge and diverse domain of communal provision and what it takes, in the way of public administrators' arrangements, to attend to the needs of its members as they collectively understand those needs. Following Walzer, engendering a sensitivity to need encourages a willingness to continue waging "the struggle against poverty (and against every other sort of neediness)."[16] Service delivery within the private sector does not go beyond immediate needs of customers. Thus, being trustworthy as a public official is not as straightforward an action as may be imagined. Learning to be trustworthy as a public official requires recognizing that being entrusted with the things valued by citizens in a specific community brings with it certain sorts of responsibilities. It requires that public officials give attention to, and develop sensitivity toward, the value individuals place on the goods they "hand over" to public officials. The public service does not merely respond to citizens' immediate needs, but encourages citizens to fulfill their responsibilities toward other members of the community to strengthen civil society as a whole. Such sensitivity seems to be essential if public officials are to be able to understand and thus be able to provide appropriate care for the goods that citizens entrust to them.

Applying a communitarian approach to trustworthiness in public administration should be based on the communal role of public officials. Such an approach is important not only because this contributes to the flourishing of communities but also because it adds to the general perception of the trustworthiness of public administration as an institution. In addition, by being trustworthy, public officials contribute to a general climate of trust and this is essential if trust itself is to flourish. However, the ethical weight of the communal principle of trustworthiness should be restricted by impartiality when attachments and partialities cause the public official to become so emotionally close to any group that he does not handle or dismisses the interests of other groups.

Maximizing Community Benefits and Reducing Harms

Walzer is attempting to give an account of the nature of distributive justice as we already intuitively understand it, or, failing that, as we could be brought to understand it if we took a rational and sustained interest in doing so. Walzer is trying to understand the complex structure latent in the ideas we have long possessed of equality.

Egalitarian theories of distributive justice are almost always concerned with the assumption that differences in human skills and qualities lead to acquiring different goods in varying degrees. Part of the natural order, those endowed with skills, knowledge, and highly innovative and competitive capabilities in exchange should become wealthy, those skilled in the art of rhetoric and persuasion should gain political office, and the intellectually endowed should acquire degrees and respect. This order may not be interrupted unless one type of social good, such as wealth, begins to operate outside its bounded "sphere" and is used to acquire other types of goods. Thus, some good, or set of goods, serves as means of domination. And it is from this type of domination—whether of the poor by the rich, of the commoner by the landed aristocrat, or of the ordinary citizen by the bureaucrat—that inequality is multiplied and reproduced. Thus, Walzer has set his theory of distributive justice on the notion of "complex equality" rather than the simple form of equality that concerns the possession of one's natural qualities. Possessing goods based on the inevitable connection between one's natural qualities and dominating others by means of those goods, is discussed by Walzer:

> It is not the fact that there are rich and poor that generates egalitarian politics but the fact that the rich "grind the faces of the poor" impose their poverty upon them, command their deferential behavior. Similarly, it's not the existence of aristocrats and commoners or of office holders and ordinary citizens (and certainly not the existence of different races or sexes) that produces the popular demand for the abolition of social and political difference; it's what aristocrats do to commoners, what office holders do to ordinary citizens, what people with power do to those without it.[17]

It is under that condition of domination through a set of social goods that people seek to eliminate, if not all distribution differences, at least a particular set of distribution differences, namely that one through which domination takes place. When a set of social goods is used by one group to dominate others, human beings acknowledge the meaning of "equality," seeking to eliminate differential distributions (if not the qualities that led to them).[18]

This view of complex equality is developed by Walzer to divide the spheres of control so that no good, or set of goods, determines value for all other goods. Walzer then examines the ways in which a dominant good or set of goods has varied across various communities and historical periods. For example, in states under a feudal regime, the dominant good was privilege; in modern market or industrial societies, it is capital; in autocracies and oligarchies, it is power. In each society, a group of individuals gains monopoly over the possession of a dominant good, which paves the way to accumulate all other types of social goods, even when those other social goods are subject to other terms of distribution and distributive principles, ones in keeping with the intrinsic meanings of the goods in question. Viewed in this way, monopoly of a dominant social good leads to social and political domination, through which moral and political inequalities are multiplied. This process is described by Walzer:

> Most societies are organized on what we might think of as a social version of the gold standard: one good or set of goods is dominant and determinative of value in all the spheres of distribution. And that good or set of goods is commonly monopolized, its value upheld by the strength and cohesion of its owners. I call a good dominant if the individuals who have it, because they have it, can command a wide range of other goods. It is monopolized whenever a single man or woman, a monarch in the world of value—or a group of men and women, oligarchs—successfully hold it against all rivals. Dominance describes a way of using social goods that isn't limited by their intrinsic meanings or that shapes those meanings in its own image. Monopoly describes a way of owning or controlling social goods in order to exploit their dominance. . . . Physical strength, familial reputation, religious or political office, landed wealth, capital, technical knowledge: each of these, in different historical periods, has been dominant; and each of them has been monopolized by some group of men and women. And then all good things come to those who have the one best thing. Possess that one, and the others come in train.[19]

Thus, Walzer's egalitarian mission is to break the monopoly of a dominant good or a set of goods by ensuring a relative (local) autonomy among them: within each sphere one can identify a distinctive distributive principle following from the nature of the social good or goods around which the sphere is constituted.[20] The critical power of the idea of local autonomy is made explicit by the nondominance principle: "No social good x should be distributed to men and women who possess some other good y merely because they possess y and without regard to the meaning of x."[21] Use of the local autonomy construct underlies the complexity inherent in the idea of equality. While the egalitarian enterprise attempts to set up powerful state systems that ensure equal, or nearly equal, shares to everyone in order to break the monopoly of a dominant good, based on only one distributive criterion (e.g., free exchange, desert, need, utility), Walzer's complex egalitarian approach leads to development of the "pluralism of distributive possibilities," which is an acceptable and complementary system of distributive justice that he characterizes as "egalitarianism that is consistent with liberty."[22] Starting with the premise that we live in a pluralistic society, allocation of goods will be achieved according to the principles of "different goods to different companies of men and women for different reasons and in accordance with different procedures."[23] The protection of equality rests on the idea that inequality in the holdings of one particular good should not translate into inequality of other social goods.[24]

Thus, the legitimacy of complex equality, as a just system of distribution, depends heavily on establishment and maintenance of shared understandings among community members that are dynamic and tenuous in nature:

> The sharing of sensibilities and intuitions among the members of a historical community is a fact of life. Sometimes political and historical communities don't coincide, and there may well be a growing number of states today where sensibilities and intuitions aren't readily shared; the sharing takes place in smaller units . . . But this adjustment must itself be worked out politically, and its precise character will depend upon understandings shared among the citizens about the value of cultural diversity, local autonomy and so on. It is to these understandings that we must appeal when we make our arguments—all of us, not philosophers alone; for in matters of morality, argument simply is the appeal to common meanings.[25]

Thus, for individuals who possess skills and knowledge competing in the sphere of office, only the office would be at stake, and nothing more. Attainment of an office would not carry with it honor, power, wealth, or goods from other spheres.[26] Walzer then provocatively asserts that it is all right to be "unfair" in the distribution of some goods, as long as these inequalities are bounded to their distributive sphere and balanced by other inequalities existing in other spheres.[27]

In searching for the principles that systematize and explain shared moral judgments and intuitions that maximize benefits to socially disadvantaged groups ultimately benefits society as a whole, Walzer articulates six basic propositions.[28]

1. The meaning and value of all goods (money, power, love, and so forth) are socially defined and vary from one society to the

other. Social goods do not include privately valued goods, such as sunsets or mountain air.[29]
2. Self-identities are created by the way goods are conceived, created, and exchanged in certain manner by those within a society.[30]
3. No one good or set of goods dominates other goods as universally accepted, or most important; even the possible range of goods perceived as needs is very wide.[31]
4. The meanings of goods are local in nature; thus, all distributions are just or unjust relative to the social meanings of the goods at stake.[32]
5. The social meanings of goods are dynamic in character; thus, the extent to which distributions are considered just or unjust changes over time.[33]
6. Every social good or set of goods constitutes a distributive "sphere," which is characterized by a particular set of principles or rules to guide the way in which these goods are distributed, shared, or exchanged.

Public officials should advocate for distribution of welfare promotion and harm-prevention resources in ways that are fair and equitable in benefiting the most people, while also considering the needs of vulnerable populations. The complex equality form of distributive justice creates the value of the local context—which I argue is the value that gives administrators their unique character among mediating institutions, which Walzer calls "professional helpers."[34] As viewed by Walzer, the local judge, for example, can support a claim for the advantage of this kind of intimate connection to a particular group of people while complying with efficacy and the office rules:

> The local judge, the connected critic, who earns his authority, or fails to do so, by arguing with his fellows . . . his appeal is to local or localized principles; if he has picked up new ideas on his travels, he tries to connect them to the local culture, building on his own intimate knowledge.[35]

To my mind, Walzer's theory represents an attractive qualification of the ideal need for public administrators to use communal settings wherein they can form their local "intimate knowledge" such as child-care centers, schools, churches, and neighborhoods, to generate information and knowledge of distributive goods internally valued as needs by their recipients.

I have now outlined the main tenets of Walzer's communitarian view of distributive justice. Walzer's theory brings attention to the role of community membership and communal provisions that are particularly relevant in light of the central role community occupies in disaster programs and policies. Following Walzer's theory, reason is always internal to bounded political community and so always internal to social meanings. Since the resilience approach to community-based emergency management resides in the active interactions between a community and aspects of the environment it experiences, it does not bear the danger of blaming the individual, the victim of the disaster event, rendering that person personally responsible for his or her problems. It is the community that establishes the form and shape as well as the circumstances and the background for resilience. Thus, it is useful to apply Walzer's communitarian theory of social justice to analyze the equity of resilience doctrine, with concern for the interrelationship of institutions such as public administration and affected communities. In the following section I address the communitarian ethical grounding set by Walzer to the central guiding principles of the resilience management doctrine, including advocacy, inclusion, and competency.

Advocacy

Advocacy centers on the relationship between a group or organization providing advocacy—the advocate—and the community being supported—the partner. In 2002, Action for Advocacy defined advocacy as:

> Taking action to help people say what they want, secure their rights, represent their interests and obtain services they need. Advocates and advocacy schemes work in partnership with the people they support and take their side. Advocacy promotes social inclusion, equality and social justice.[36]

Merriam-Webster's Collegiate Dictionary[37] defines advocacy as "the act or process of advocating or supporting," and an advocate as "one that pleads the cause of another." The notion of advocacy gained recognition in health-care system delivery during the late 1970s, during which patient advocacy became increasingly important. Early use of advocacy can be traced back to the ethical aspect emphasized by Florence Nightingale to encourage patients to act on their own behalf.[38] The ethics of the nursing profession recognized this principle in guiding nurses' training and practice by initiating the Code of Ethics for Nurses with Interpretive Statements by the American Nurses Association (ANA) requiring that nurses advocate for, and protect the health, well-being, safety, values, and rights of patients in the health-care system.[39]

Curtin maintains that the relational aspect of advocacy is the basis of the nurse–patient relationship.[40] The professional role of nurses expresses advocacy by creating an environment that is open and supportive of decision making. To apply this principle, nurses must acquire a sustained and intimate knowledge of the person as a distinct human being.[41] There are a number of key principles that are central to the type of relationship and environment evolving around the concept of advocacy: support the patient as partner in articulating his or her views; motivate the expression of concerns and opinions on behalf of the patient while remaining neutral; raise any issues where required but only if the patient wishes for that; allow the patient to make his or her own decisions and choices; ensure that a patient retains control over his or her life domains; develop a one-to-one relationship with the patient; understand and respect the rights of patient confidentiality.

In addressing advocacy and collaboration to the community-helper relationship, a specific awareness should be paid to the particular difficulties in the environment.[42] Thus, to work with communities to bring about advocacy, Lewis et al. stress the need to train helpers in gaining skills in interpersonal relations, communication, and research, which can be made available in engagement and collaboration with communities.[43]

Advocacy at the community level is formed by the following eight qualities required of those who work with communities:

1. identify environmental factors associated with community development;
2. inform community with common concerns related to the issue;

3. develop partnerships with groups working to bring change and competence;
4. develop effective listening skills to gain understanding of the group's goals;
5. recognize the strengths and resources that the community members possess in the process of systemic change;
6. motivate and communicate recognition of and respect for these strengths and resources;
7. recognize the skills that the helper can bring to the collaboration; and
8. assess the effect of the helper's interaction with the community.

Lewis et al. acknowledged the need to overcome emotional and practical barriers such as "vision, persistence, leadership, collaboration, systems analysis, and strong data."[44] Thus, those who work with communities, by virtue of their training, possess these qualities with a strong professional commitment as part of their professional roles to bring about change in their clients.

The principle of advocacy places greater importance on local knowledge, including understanding the practices of communities at risk that better mobilize these communities in the pursuit of collective goals. This principle suggests that public officials should respect this knowledge in policy deliberation and practice and create opportunities for affected communities to develop their capabilities.[45]

In applying Walzer's theory of justice, it is claimed that there is a plurality of goods found in spheres such as religion, kinship, law, politics, economics, aesthetics, and education; however, each good or cluster of goods has its own social meaning, which shapes particular patterns of division and exchange. Thus, goods are always social in nature in that they are objects of exchange, both physically and symbolically, including human practices, traditions, customs, institutions, and ways of life that structure and shape so much of our experience.[46]

Therefore, administrative intervention programs should follow Walzer's idea of public administrators as professional "helpers," as their aim is not the provision of goods but rather as partners that support the partners' active engagement in coordination and distribution decision-making with their goods providers for positive outcomes. "By virtue of what characteristics are we one another's equals? One characteristic above all is central to my argument. We are (all of us) culture-producing creatures; we make and inhabit meaningful worlds."[47] Public administrators are then required to become attentive to the shared cultural identity that establishes a "we" as a prerequisite for ensuring democratic equality in terms of positive experience with political participation.

Competency

The term *competency* refers to a state or quality of being able and fit. A competency is understood to be more than just knowledge and skills. It involves the ability to meet complex demands by drawing on and mobilizing psychosocial capacities (including skills and attitudes) in a particular setting. For example, the ability to communicate effectively is a competency that may have great bearing on an individual's knowledge of language, skills, and attitudes and commitment toward those with whom he or she is communicating. The definition of competency underlies three broad categories of key competencies. The first category refers to the ability of individuals to use a wide range of tools for interacting effectively with the environment. These include technology and sociocultural tools to adapt them for their own purposes—to use tools interactively. The second category is associated with the ability to communicate and interact within heterogeneous groups and individuals from a range of backgrounds. Third, individuals need to be able to take responsibility for managing their own lives autonomously.

Within the field of human resource management and job performance competencies are references to collections of knowledge, skills, abilities, and other characteristics (KSAOs) that are needed for effective performance in the jobs in question.[48]

However, this conceptualization is more associated with directly observable and testable competencies, such as knowledge and skills, while less assessable competencies are related to personal characteristics or personal competencies. Although the unobservable competencies have not yielded in-depth investigations, some scholars recognized the role of such competencies. For example, Boyatzis defines competency as "an underlying characteristic of a person which results in effective and/or superior performance in a job."[49] Accordingly, underlying characteristics could include a motive, trait, skill, an aspect of one's self-image or social role, or a body of knowledge. Spencer and Spencer, who followed Boyatzis's work, use competency to denote the "underlying characteristic of an individual that is causally related to [a] criterion referenced [as] effective and/or superior performance in a job or situation."[50] McClelland has suggested visualizing competencies as an iceberg with a person's knowledge and skills representing the visible tip of the iceberg, while the underlying and enduring personal characteristics or self-concepts, traits, and motives, such as self-confidence, initiative, empathy, and achievement orientation, represent the larger portion of the iceberg, hidden below the waterline.[51] The constant changes in today's world lead to a decrease in the "shelf-life" of knowledge and skills; thus the long-enduring, "below the water-line" competencies have a more substantive impact on how effectively an individual performs on the job.

In sum, competencies are suited to indicate ways of behaving or thinking, which generalize across a wide range of situations and endure for long periods of time, and include five key components: knowledge, the information and learning residing in a person; skills, which concern a person's ability to perform a certain task; personal concepts and values, associated with a person's attitudes, values, and self-image that he or she can carry out a given task; physical traits refer to physical capacities and consistent responses to situations or information; motives refer to emotions, desires, physiological needs, or similar impulses; and interpersonal orientation needed to perform responsibly toward tasks and other members of the workplace environment.[52]

Resilience is based on the principle that all individuals, families, and communities have some control and/or competencies over their environment, and as recipients of help are not totally dysfunctional. Thus, people can exercise their strength and competence in their lives through their personal interaction in various social institutions such as families, neighborhoods, churches, and voluntary associations, and these institutions will be best at providing effective assistance based on a more balanced exchange.

Recognizing the pluralist nature of society and personal relationships, Walzer proceeds to describe a system of distributive justice that he characterizes as "egalitarianism that is consistent with liberty."[53] Starting with the premise that we live in a pluralistic society, allocation of goods will be achieved according to the principles of "different goods to different companies of men and women for different reasons and in accordance with different procedures."54 For example, in the sphere of welfare and security, the prevailing distributive principle is of "providing according to need."

Walzer's complex equality provides the need to address local understanding—which I argue is the value that gives administrators their unique character among mediating institutions, or as Walzer calls them, "professional helpers."55 According to Walzer, public administrators need to enhance their communicative skills so they will be able to form their local "intimate knowledge" provided by small, intimate social institutions, such as child-care centers, schools, churches, and neighborhood, where the strength and the competence of the community are reflected, to generate information and knowledge of distributive goods, internally valued as need for their recipients. It is claimed that participating in local institutions reflects community members' judgment about whether they are capable of carrying out the social tasks that underlie successful relations with others.

Inclusion

Inclusion, or the act of being included, means being accepted and able to participate fully within the family, the community, and the society within which one lives.56 People who are excluded, whether because of poverty, ill-health, gender, race, or lack of education created by disaster effects, do not have the opportunity for full participation in the social and economic benefits of the community or the society. Social inclusion denotes the need to accept someone into interpersonal interactions and social networks. Thus, considerable attention has been paid to defining social exclusion as the opposite of social inclusion. The term *social exclusion* was first coined in France in 1974, and since then has gained increasing popularity and usage in different contexts and in a range of academic, political, and professional arenas. Social exclusion was offered to identify an underclass that fell outside the protection of the state's social insurance. These groups were labeled "mentally and physically handicapped, suicidal people, aged invalids, abused children, substance users, delinquents, single parents, multi-problem households, marginal asocial persons, and other social 'misfits.'"57

Within political, academic, and professional discourses, social exclusion is often associated with words such as *poverty* and *deprivation*, which confine it largely to the context of social and economic policy. However, the growing use of social exclusion, instead of poverty or deprivation, in the area of policy analysis and decision making is due largely to the recognition that poverty has been seen as too narrow and limiting a concept in addressing the social problems.[58] The poverty/social exclusion debate reflects the fundamental differences in the ways in which different societies view inequality and disadvantage. Room[59] proposed viewing poverty in terms of allocation and distribution of resources, whereas social exclusion is concerned with relational issues (emphasizing the relations between an individual and various potential support networks, such as family, friends, community, and state services and institutions).

By focusing on relational issues, social exclusion is conceptualized on a multidimensional basis, incorporating economic, social, and political dimensions. Bhalla and Lapeyre[60] identify these three interrelated dimensions of social exclusion for "it is useful to demonstrate that political freedom and civil rights and liberties can draw the best out of people and raise their productivity, thereby contributing to growth and overcoming economic exclusion."[61] The economic dimension is associated with income and production issues as well as access to goods and services: social deprivation thus results from "income and livelihood, and from the satisfaction of such basic needs as housing/shelter, health and education."[62] The social dimension of social deprivation is concerned with social resources and competencies such as self-identity, dignity, and the role of community and other supportive networks that empower citizens to obtain substantial influence on decisions that affect them. Bhalla and Lapeyre characterize the social dimension as incorporating access to social services (e.g., health and education), access to the labor market, and the opportunity for social participation and its effects on the social integration. The political dimension of exclusion includes access to various rights, such as civil (e.g., the right to justice, freedom of expression), political ("the right to participate in the exercise of political power"), and socioeconomic (equality of opportunity, right to minimum welfare benefits, etc.) rights.

It is then argued that how communities interpret and deal with exclusion and stigma is an important factor in building community resilience, since discrimination and social exclusion processes affect communities' mental and physical resources and well-being. Walzer's view of a community of justice seems to offer limited inclusiveness as it is always bounded by dense webs of common understandings or shared social meanings. The boundaries of the political community entail obligations we have to other members. Obligations flow from belonging, identity, and relatedness. Thus, the moral role of memberships in certain constitutive communities, or constitutive ties more generally, can be an important variable in advancing a sense of mutual care and obligation: "Morality takes shape as a conversation with particular other people, our relatives, friends and neighbors."[63] Thus, relations of reciprocity and mutual trust generate the necessary solidarity, willingness to help others, and so to act responsibly.[64]

However, the moral effects derive from engagement in a community of justice where special, stronger obligations to others, formed by shared history or close personal relationships, can serve as a channel for extending obligations beyond the boundary of the political community and proceed to those more geographically or culturally distant from us ("society of equals"). The applicability of this inclusive view of community is actualized in the sphere of security and welfare, where goods are to be collectively provided to each needy person in proportion to his or her neediness.[65]

Walzer's theory thus provides a form of inclusivity that is based on responsibility and reciprocity that has self-determination, rather than coercion, at its core, which offers potential safeguards to communities' substantive capacity to choose, and to their potential for empowerment and active participation.[66]

In summary, the process of building resilient communities is based on normative assessment of four main government and administrative efforts aimed at improving the capacity of disaster-affected communities to cope with emergency circumstances. This chapter suggests that issues surrounding the management of disaster resilience involve a communitarian framework of social justice. In our particular case, in affected communities organized around a common good containing a commitment to improve members' capacities and resources, it is likely that needs-meeting is functional for resilience policy in natural disaster events.

It appears that Walzer's approach to social justice deals with community members as we find them—that is, people standing in relation to a set of goods, with a history of transactions. Walzer's concept of community of justice offers to show that distributing government resources according to need across affected communities contributes to sustaining people's lives, so that the relation between justice and resilience management would be one of reinforcement. Moreover, it seems that Walzer's principle of need is compatible with increasing the knowledge of community valuable goods and needs, so long as it is functional for securing community-valued goods. Following Walzer, the respect for persons is extended to respect for their experience as local knowledge, and their abilities and capacities as community members to contribute to decision making and planning at the community level. Taking on Walzer's concept of community, nurturing relations of reciprocity, mutual trust, and so forth in communities generates the necessary solidarity, willingness to help others, and so to act responsibly and promote inclusion.

Thus, resilience disaster management should be constructed so as to give space for community-administration interaction as formalized by Walzer's theory of justice. Indeed, public officials are often uniquely situated to respond to someone's need, which derives from an ongoing relationship with the person by virtue of which one has been held responsible for his or her well-being. In this sense, the Walzerian theory of social justice is morally instrumental in a relationship to the extent that it contributes to the protection of those who are in need and assigns responsibilities for the care of needy persons to others who stand in certain relationships to them.

The ethical evaluation of the extent to which public administration and community interactions in different regional settings enhance the disaster resilience of affected communities by meeting the communitarian justice approach set down by Walzer will be further expanded in the next chapter.

6
Comparative Analysis of Community-Based Disaster Resilience Policies

This chapter evaluates the principles underpinning a resilient community using Walzer's theory of social justice, particularly as they impact communities and public administration relations and their implications on disaster management policies in various recent contexts, including the Gulf Coast Hurricanes in the United States (2005), the West Sumatra Earthquakes in Indonesia (2004), the Great East Japan Earthquake (2011), and the Wenchuan Earthquake in China (2008).

Comparative Analysis of Disaster Resilience Policies

Let us consider how communitarian justice criteria add a new dimension to the evaluation of public administration practice in these recent disasters. Resilience plans were strongly criticized by a number of scholars who claim that such plans reflect government and bureaucratic decision-making breakdown in the face of pressing need.[1] Indeed, heated debates and concerns address racial and exclusionary bias in the inefficient management, rescue, and relief actions by public administrations in response to disaster effects.[2] Since this paper addresses a community-based disaster risk approach to society–administrative relations, the term *civil society* as a category used to distinguish between autocratic and democratic regimes, will refer to the community of a country's citizens' active engagement in the governing of the country within legally defined limits.[3] The American context represents old and mature civil society; the Japanese case represents semiindependent civil society that is often allied to or collaborates with the state; the Indonesian case represents young civil society, which until 1998 experienced dependence on an authoritarian regime; and the Chinese case represents a greatly dependent civil society in an autocratic state. Thus, the comparison between these cases will also provide insights into the role of public administration in disaster emergency management as part of the process of building civil society capacity.

Natural disaster crises prompted a variety of programs and services to address individual and community issues. This section will review some regional programs that utilize resilience strategies, including the Gulf Coast Hurricanes (United States), the West Sumatra Earthquakes (Indonesia), the Wenchuan Earthquake (China), and the Great East Japan Earthquake; resilience plans that were strongly criticized by a number of scholars who claim that such plans reflect government and bureaucratic decision-making breakdown in the face of pressing need.[4] It should be noted that some of these programs were designed using the resilience principles as guidelines, while others arose naturally in response to concerns and issues. Indeed, heated debates and concerns address racial and exclusionary bias in the inefficient and inequitable management, rescue, and relief actions by public officials in response to disaster effects.[5]

On 26 December 2004, Indonesia experienced the 2004 Sumatran earthquake. About 170,000 people were reported dead, with an estimate of more than 37,000 missing. In addition, it was reported that hundreds of buildings had collapsed, which left thousands of people homeless.[6] On August 29, 2005, the center of Hurricane Katrina passed east of New Orleans; winds downtown were in the Category 3 range, with frequent intense gusts and tidal surges. At least 1,836 people lost their lives and 80 percent of New Orleans was flooded, with some parts under 15 feet (4.5 meters) of water.[7] Another deadly disaster occurred on May 12, 2008, in the Sichuan province of China (also known as the Wenchuan Earthquake). At least 69,000 people were killed, 374,176 were injured, with 18,222 listed as missing; about 4.8 million people were left homeless.[8] A more recent case that is dealt with in this book is the Tōhoku earthquake and tsunami that occurred in Japan on March 11, 2011. The Japanese National Police Agency has reported 15,760 deaths, 5,927 injured, and 4,282 people missing. In addition, over 125,000 buildings were damaged or destroyed. The disaster led to nuclear accidents at three reactors in the Fukushima I Nuclear Power Plant complex, triggering a thirty-kilometer evacuation zone surrounding the plant. By studying these examples, it is possible to examine the extent to which disaster governmental and administrative deliberations in disaster relief efforts in these regions meet resilience principles, their degree of success, and the limits of these efforts.

Disaster Resilience Management in the United States: The Gulf Coast Hurricanes, 2005

Civil Service of United States

The civil service was established in 1872. The Jackson administration used the spoils system (also known as the patronage system) to encourage voters after a political party wins an election by providing government offices and appointments to its voters as an incentive to ensure that the party's voters and supporters keep working for the party—as opposed to a system based on merit. During and after Jackson's presidency, the federal bureaucracy grew from almost 20,000 to over 120,000 federal employees in 1883. This practice was gradually challenged by the introduction of the Pendleton Civil Service Reform Act of 1883 and subsequent laws. The Pendleton Civil Service Reform Act was part of the civil service reform movement, which laid the groundwork for merit-based civil service.[9] This act transformed the nature of public service for the years to come. For example, in 1909 almost 70 percent of the federal employees were appointed based on merit. Under the Pendleton Civil Service Reform Act, the United States Civil Service Commission was established to administer the civil service of the United States federal government.[10] The law set up merit-based criteria in recruiting federal government employees and protecting civil servants from the influences of political patronage and engagement in partisan political activity. A further move in diminishing political patronage in government jobs was laid by the Hatch Act of 1939, which prohibited civil servants from engaging in political activities while performing their duties.[11]

The Federal Civil Service is defined as "all appointive positions in the executive, judicial, and legislative branches of the Government of the United States, except positions in the uniformed services."[12] The civil service is divided into three categories: the Competitive Service, the Excepted Service, and the Senior Executive Service. These categories are differentiated by the areas of appointment procedures and job protections. In competitive service, the recruitment procedure is based on merit promotion requirements; qualification requirements are outlined by law or by the Office of Personnel Management and apply to all agencies. The Office of Personnel Management (OPM) is the federal government human resource agency that works with the president, Congress, departments, and agencies, and other stakeholders to connect job applicants with federal agencies and departments.[13] OPM connects people according to their expertise and skills to meet the specific needs of federal agencies to maintain successful, high-performance organizations. In the excepted service, the recruitment procedure follows basic requirements that are set by law or regulation and each agency outlines its own specific requirements and procedures for its own jobs. Under the executive category are presidential staff organizations including the White House staff, the National Security Council, the Office of Management and Budget, the Council of Economic Advisers, the Office of the U.S. Trade Representative, the Office of National Drug Control Policy, and the Office of Science and Technology Policy.

Today, the majority of civil service appointments in the United States are made under the Competitive Service, while in agencies such as the Diplomatic Service, the FBI, and other National Security organizations recruitment is made under the Excepted Service.[14] Figures have shown that as of January 2007, the federal government, excluding the Postal Service, employed about 1.8 million civilian workers. The federal government is the nation's largest employer (over 1,300 federal government agencies). Most federal agencies are located in the Washington, DC region, while fewer than 20 percent of the federal government workforce is employed there.[15]

Advocacy

The experience of Katrina indicates some violations of the right to adequate housing, health, and dignity, which are viewed as part of a broader pattern in the U.S. social and urban policy in claiming a "war on the poor" involving the redistribution of resources and opportunities downward from low- and middle-income people toward the most vulnerable communities of American society.[16] The right to an "adequate standard of living," which includes the right to housing, acknowledges the fundamental need that people have for safety and security, as articulated by the Universal Declaration of Human Rights (UDHR) and the International Covenant on Social and Economic Rights (ICESCR), which have been most widely adopted. This recognition draws on government's obligation to ensure access to adequate housing for everyone, with an emphasis on the most vulnerable.[17]

Viewed in this way, Katrina's impacts on the people of New Orleans reveal public authorities' goal to build a "new" New Orleans from the ruins left by Katrina's passage. On September 15, 2005, President George W. Bush addressed this mission as the need to rebuild a "better and stronger" Gulf Coast through a "Gulf Opportunity Zone," suggesting that "it is entrepreneurship that creates jobs and opportunity . . . it is entrepreneurship that helps break the cycle of poverty." Bush stated that his administration would "take the side of entrepreneurs as they lead the economic revival of the Gulf Zone."[18] It should be noted that for administrative agencies to be entrepreneurs that create public value such as the fight against poverty, from their operations, they need to search, define, and be able to reach over agencies such as civil society organizations that are eager to share their valuable resources in disaster relief and reconstruction projects. Consequently, the role of administration in such a system cannot be that of a mere bearer of information and resources. Rather, public officials need to be critical value creators during disaster relief and reconstruction efforts that interprets the needs and competency of the multiple participants engaged in emergency management system and coordinates the scattered efforts of agencies to reduce the vulnerability of affected communities. In this regard, developing an administrative entrepreneurial governance network requires that public officials mediate the flow of resources and information within the reconstruction system, recognizing the discrepancies among community members. Indeed, the perceived purpose of this mission seems to use the natural disaster as an opportunity to remove the city's poor inhabitants by violating the right to housing for New Orleans's poor Black residents, who were more likely to ascribe blame for discrimination and racism to the implementation of post-Katrina recovery policies of the Housing Authority of New Orleans (HANO). The perceived powerlessness of these residents complicated their capacity to articulate their shared interests in evaluating housing policy options.[19] According to Landphair: "Amid municipal neglect and increasing impoverishment, Lower Ninth Ward residents developed cross-generational neighborhood bonds that encouraged activist pursuit of better public services and nourished cultural traditions singular to New Orleans."[20]

In this study we suggest that helping New Orleans's poor residents to better articulate their shared interests is very important for public officials when acting as communities' advocates in the case of urban renewal policies in post-Katrina New Orleans. If administrators were more attentive to local and shared meanings of house or "home" behind residents' perceptions in this particular situation, communities' advocacy actions would be meaningful, appropriate responses would be provided, and survival needs satisfied. The concept of the right to housing involves the recognition that housing is more than a safe place, but rather a place marked by familiar locals, social networks, and cultural norms and practices. Thus, housing has a deeper sense in terms of community, belonging, and reciprocity. The meaning of attachment to homeland or home ground could be revealed by communicating with survivors and/or their families in order to understand their best interests and priorities in recovery from natural disaster. For example, in a survey conducted by the NESRI/Mayday in 2010, one New Orleans resident admitted that

> When people ask me questions about my story, the first thing I talk about is the loss of my community and networks. People don't understand my pain . . . Losing my community was a heart-breaker for me. There are so many families that can't get back. In our hearts, all of us want to go back. . . . [In St. Bernard] we weren't just a community—we were a family. My neighbors would watch out for me and my children and make sure they were okay. I don't have that now. I'm on my own.[21]

The concept of housing is signified as a relational concept but it also should be defined across multiple spheres, to avoid narrow definitions that can convey a partial picture. It has been suggested that violations of the right to housing have a direct impact on health. Scholars have long stressed the correlation between deficiencies in housing and community infrastructure and health inequities among disadvantaged groups, underscoring the role of underlying social and political power relationships in terms of access to adequate housing and safe living environments.[22] Within the context of natural disasters, studies have shown that housing policy choices can effect a range of personal and community health outcomes experienced by adults, including psychological trauma and stress as well as other negative effects on children's early physical and cognitive development.[23]

In a 2008 survey conducted by the Kaiser Family Foundation to address the impacts of the Katrina storm and its aftermath on self-reported physical and mental health across communities, 12 percent of Black residents and 15 percent of the "economically disadvantaged" admitted that their mental health had worsened, as opposed to 4 percent of White residents and 4 percent of wealthier demographic groups.[24] These findings were also confirmed by the NESRI/Mayday survey, which showed that "Among survey respondents who reported that their health has drastically deteriorated since the storm, the majority cite the stress of finding housing as a major factor."[25]

Because most of the factors associated with health disparities are beyond any individual's ability to control, an advocacy approach that addresses housing policy is needed to achieve the structural changes aimed at creating healthy communities. Although poor communities in New Orleans face difficulties against government authorities and reforms, some advocacy efforts to press the needs for adequate housing and to spell out the challenges faced by the city's poor communities, in particular residents and former residents of public housing, and to articulate a set of policy demands were evident. For example, in 2006 the Rapid Evaluation and Action for Community Health in Louisiana (REACH-LA) project was initiated, aimed at identifying the needs, existing resources, gaps, and solutions to ensuring health care in New Orleans after Hurricane Katrina. This NGO's project differs from other studies of post-hurricane health issues since it used community-based participatory methods to involve community members themselves in the interpretation of the results. The project was accompanied by a Scientific Advisory Board to provide scientific oversight, and methodological and conceptual support. The information was gathered from interviews of key informants (policy makers, health-sector recovery planners, healthy system administrators, health-care providers, and community health leaders), four Community Discussion Groups (CDG) across New Orleans neighborhoods, and a Community Feedback Conference. The significance of such initiative lies in its ability to capture grassroots community perspectives to identify health-care needs in response to disaster.[26]

Among efforts to advocate for addressing equity in recovery and rebuilding processes such as the Broadmoor Improvement Association[27] and the Lower 9th Ward Neighborhood Empowerment Network Association, is the Churches Supporting Churches (CSC).[28] The Churches Supporting Churches is a coalition of national and local churches aimed at increasing the engagement of community low-income residents in policy advocacy by using participatory and formative evaluation and feeding back the results to the city recovery management officials and community members to incorporate the results into post-hurricane rebuilding programs. Viewed in this way, the CSC helped pastors and members become policy advocates to identify the needs and rebuilding status of selected neighborhoods using community-enhanced geographic information system (GIS) methods. This method was implemented through the collaboration between CSC, PolicyLink, and the Louisiana State University CADGIS laboratory to map five sixteen-block neighborhoods to provide current, ground-level data to facilitate the redevelopment of those areas.

As a technique providing visual imagery of buildings with their associated locations in space, data extracted from the Spatial Video Acquisition System (SVAS) can assist in evaluation of rebuilding, of abandonment, and the degree of damage, and to determine aggregate changes in the built environment. In 2007, five areas relevant to CSC goals were mapped to evaluate their redevelopment priorities and aspirations together with the efforts envisioned for the target development areas managed by the Office of Recovery Development Administration (ORDA). Community members were able to make informed suggestions to local and state government officials by interpreting the data and providing their input for planning equitable rebuilding programs. In this way, CSC provided the necessary spiritual connection and support to motivate community members to effectively advocate for policy decisions that advance their priorities and needs.

In 2005, a group called Mayday New Orleans was initiated by local public housing residents as an advocacy group against the demolitions of public housing in New Orleans. Mayday New Orleans aimed to fight for their housing rights through the "Stop the Demolitions Campaign" by challenging the urban redevelopment model advocated by the government and some private organizations.[29]

The advocacy strategy adopted by Mayday New Orleans involves engagement and mobilization within the local community members to identify problems, rank priorities, disseminate information, and create partnership with outside organizations to gain technical and legal support and facilitate political lobbying efforts. At the national level, Mayday engages in the Campaign to Restore National Housing Rights, a coalition of housing rights groups from around the country that have united to force the U.S. federal government to recognize its obligation to adequate housing for all.

The strategic plan developed by Mayday and its partners have shown effectiveness in strengthening and extending community voices in policy making when they were able to broker a fact-finding mission to New Orleans by the international Advisory Group on Forced Evictions (AGFE), which undertook a five-day mission to personally meet members from affected communities across New Orleans and to motivate residents to share their experiences of forced eviction, and their interests and beliefs for a more equitable rebuilding process.[30]

The fight for security and housing rights undertaken by groups like Mayday New Orleans embodies challenges associated with the use of a community-based approach for policy advocacy, including the length of time it can take, the difficulty in changing state and national policies, and the difficulty of sustaining and monitoring action for the long run. It is suggested that NGOs rather than administrative agencies became entrepreneurs that were able to identify community's needs in disaster relief and reconstruction efforts. The role of public administration in such a system was to join and collaborate in NGOs' collaborations and activities. It is suggested that these challenges make the study on community-public administration partnership for policy advocacy all the more important. Public administrators should act as community advocates. It is important for administrators to motivate community members to take actions to advocate for themselves, but in cases where members are incompetent or want administrators or someone else to represent them, public officials should act on their behalf, representing their needs and best interests. Appropriate knowledge and skills related to community advocacy, education, or training is necessary since administrators may need to know how to negotiate between communities, NGOs, and other care providers (e.g., physicians, local businesses). It is crucial for administrators to identify barriers to advocacy policy such as a lack of support from institutions, government, or both.

Thus, according to our proposed model, community advocacy is viewed as a process consisting of a series of specific actions for preserving, representing, and/or safeguarding community members' rights, best interests, and values in disaster management. Public administrators should be able to identify the kinds of situations in which communities need an advocate, what communities' best interests are in a particular situation, and what kind of actions need to be taken to preserve, represent, and/or safeguard communities.

Inclusion

Inclusion initiatives are often crucial for racial and ethnic minorities because they affect their ability to recover from disaster. There is growing recognition that community inclusion is a viable approach for addressing racial and ethnic disparities and poverty, and that racial and ethnic communities can actively participate in the public sphere to affect the policy change process in communities aimed at eliminating such disparities.

It is that kind of model of participation, that kind of focus on the collective and the social, which provides a model of responsibility and reciprocity suitable to a form of inclusivity that has self-determination rather than coercion at its core. Therefore, that model of citizenship and reciprocity provides potential protections to community members' substantive capacity to choose, and to their potential for empowered expression and participation in recovery decisions. In the aftermath of Hurricane Katrina, several organizations paved the way for inclusive processes essential to ensure that the Gulf Coast rebuilding effort does not intensify historical and systematic social and racial inequities that were exposed by the high levels of poverty in affected, primarily ethnic Gulf Coast communities. This process aims to provide a place at the rebuilding/reconstruction table to prevent the exclusion of people of color from rebuilt areas. The New Orleans Coalition on Open Governance was founded in post-Katrina New Orleans to challenge the practices of exclusion from decision making that threatened the equity and sustainability of city/region. The New Orleans Coalition on Open Governance (NOCOG) consists of six groups committed to promote open, responsive, and accountable government and governance in New Orleans, including the Committee for a Better New Orleans, the Lens, Neighborhood Partnerships Network, Public Affairs Research Council, the Public Law Center, and Puentes New Orleans Inc. The NOCOG calls for community participation in civic discussions and decisions, increasing access to public data and information, using media and communications that inform and affect stakeholders, and seeking inclusive public policy process and developments.

It is argued that informed people are less vulnerable and thus can act collaboratively and deliberatively with government to create accountable and more inclusive communities.

The involvement of the Committee for a Better New Orleans (CBNO) in NOCOG provides a broad-based, diverse representation of any organization in the city and a focus on change at systemic levels.[31] CBNO engages in promoting the program of New Orleans Citizen Participation (CPP), which enables citizens to effectively participate in city government's priority-setting and decision-making, and to provide an arena for open dialogue between communities, neighborhoods, and city administration and government. This initiative is set to include the rights and needs of all communities for building a consensus-based decision-making structure that addresses the interests of the city as a whole.

The Lens was formed in 2009 as the first nonprofit journalism venture in New Orleans and serves as a reliable and authentic database of information for Gulf Coast communities.[32] As a watchdog, the purpose of the Lens is to enhance Gulf Coast communities' capacities to use proper and adequate information necessary to actively participate in decision making. The Lens is managed by professional editorial and research staff, as well as being a collaborative network of associated organizations including the Center for Public Integrity, American Public Media's Public Insight Network, Project on Government Oversight, and the emerging national Investigative News Network. Another joint venture that has operated in New Orleans since January 1988, is the Tulane Loyola Public Law Center (TPLC). This venture aims at providing legislative and administrative services to communities, including negotiation and preparation of community benefit agreements; encouraging transparency in government, including open meetings and public records reforms; collaborative planning and citizen participation in the land use planning process; support for ethics efforts, such as the Ethics Review Board and Office of Inspector General; and other activities that promote the public interest. The Center operates with staff and law students who help represent disadvantaged people in the legislative and administrative processes of government (e.g., the elderly and disabled, children and families, poor and minority populations), appearing at the legislature and before state agencies to present the bills and proposed regulations they have researched and drafted for communities.[33]

The Public Affairs Research Council of Louisiana (PAR), founded in 1950, is an unbiased source of information on state and local government in Louisiana.[34] PAR not only provides citizens with information and databases about public issues but also empowers citizens so that their voices are heard. The extensive research conducted by PAR on Louisiana state and local government through decades, monitors implementation of reforms and reminds public officials of promises made.

To enhance communication between neighborhoods, the Neighborhood Partnership Network (NPN), a nonprofit organization, functions as a citywide network of neighborhoods that was founded in the wake of the Katrina disaster. The network aims at utilizing neighborhood collaboration necessary to increase access to decision making at government level to include the voices of individuals and communities across New Orleans. This organization motivates New Orleans's citizens to become actively involved in the formal decision-making processes that impact quality-of-life issues raised in neighborhood rebuilding and civic processes for both individuals and communities.

The mission of Puentes New Orleans, a nonprofit community development organization, is much narrower in its missions than NPM, but shares with it the need to enhance community inclusion and participation in decision making.[35] Puentes New Orleans, Incorporated, was established in response to the needs of the New Orleans area Latino community in April 2007, in the aftermath of Hurricane Katrina. The Latino community leaders argued that the Latino community was experiencing systematic exclusion from city decision-making and planning processes. The 2008 State of Louisiana Legislature raised growing concerns among many immigrant Latinos when several state legislators authored discriminatory anti-immigrant bills that attacked average citizens for being good community members. Puentes quickly joined in the effort to challenge these bills with its partners, such as Common Good, the Language Access Coalition, and the Equity and Inclusion Campaign.

Thus, Puentes's initial efforts were concentrated in addressing the historical powerlessness and voicelessness of the Latino community by developing social capital, a unified Latino community-centered voice, and strong and skillful Latino community leaders. Only fifteen months after its establishment, Puentes had become a leading Latino organization in the New Orleans area. Puentes has dedicated its recognition and resources to provide education and organizational facilities to challenge anti-immigrant laws in the State of Louisiana Legislature.

In housing issues, Puentes provides homebuyer training courses in Spanish, and housing counseling and credit review services through collaboration with the Consumer Credit Counseling Services of Greater New Orleans. The vision of Puentes in this field is to become an affordable housing developer to identify potential opportunities and developing strategies for the Latino community members. In the domain of public safety, Puentes has collaborated with the Hispanic Apostolate Community Services Catholic Charities and the New Orleans Police Department, and with support from Baptist Community Ministries and the United Way of Greater New Orleans, to build a public safety program aimed at improving relations between local law enforcement and the Latino community. In the winter of 2008, Puentes started to develop a public leadership training course with support from Common Good, in which private citizens began to receive training in public participation. This leadership training program aims at creating a pool of public leaders who wish to serve on a volunteer basis with Puentes in negotiating within the community on issues that are of value to the Latino community and address them, be they local, state, or federal.

The kind of inclusionist policies evident in New Orleans education has been a major government preoccupation. Katrina created an optimistic feeling for the city's school system, which was long regarded as one of the worst in the United States.[36] Thus, the circumstances were favorable for caring reform within the poor educational system that should have been governed by choice and competition that are structured in the American educational system.[37]

Under the Bush administration, New Orleans was a fertile ground for reforms to dismantle urban education systems across the country. More than $48 million in federal money has been channeled to charter schools in New Orleans. This initiative has raised "concerns that federal officials want to fuel a charter-based system, since similar amounts have not been allocated for traditional public schools."[38]

Studies have shown that when trying to introduce school reform, increased levels of distrust among community members could result.

This could be explained by the weak network between parents and educators and school–community relations as a whole, especially in the schools serving large numbers of poor students, as prevailed before the storm. The effects of the storm, which led to disrupted mail, lack of communication, disrupted social networks, high housing costs, and increased violent crime in post-Katrina New Orleans during 2005 and early 2006 created additional burdens on the old educational system. These burdens, together with the uncertainty surrounding the return of New Orleans's population and student enrollment in particular, which serves as a key determinant for how much funding should be distributed and is available for hiring teachers and opening school buildings, reduced the possibilities for successful school reform.[39] New Orleans teachers' union officials view the charter schools reform in New Orleans as a direct attack on their collective bargaining rights. Louisiana schools did not need to negotiate with teachers' unions, while New Orleans was one of only a few school districts in the state where teachers' unions had accomplished some successes in gaining collective bargaining rights. This issue raises the implications of reducing teachers' position and authority within the education system, which may contribute to the unwillingness of teachers to assume personal responsibility for what happens in their schools. The Orleans Parish School Board voted in June 2005 not to extend its agreement with the United Teachers of New Orleans, so even teachers working in the handful of district-run schools could not be protected by a union contract. In addition, Katrina also heated previous challenges to public education in New Orleans concerned with lack of resources along with a prison-like atmosphere including discipline policies that penalize and remove students instead of providing support for them and facilitating positive development and adaptation. While this concern predated Katrina, it has been replicated in the new "network" of schools operated by local and state officials and under various charter groups.[40] The complexity of lack of resources and the failure to provide quality education, combined with punitive discipline policies, has led to criminalizing and excluding youth from traditional education settings, which has been defined as a "School-to-Prison Pipeline." This phenomenon calls teachers and administrators to fight for the inclusion of children, especially in times of postdisaster events.

Thus, in post-Katrina New Orleans, education can provide a good case for examining the negative implications of advancing recovery efforts according to need but disregarding community values and norms. Applying Walzer's theory of justice in the decisions and reforms in the education sphere of New Orleans requires promoting the inclusion of the interests of disadvantaged people such as poor parents in evaluating reform options in the educational system. Resilience should be consistent with shared and collective norms and values and demonstrate its concern for the most vulnerable residents, even at the cost of maintaining the troubled education system of the past. This adds a refreshing framework to the social inclusion landscape in terms of building trust and collaboration, rather than efficacy and control that prevails for the long run.

Thus, public administrators who pursue the common good by promoting a rejuvenated school system for bringing residents back into the city need to take into consideration the effects of uncertainty along with the traditional patterns of interaction between parents, students, and community with the school system, as these factors are likely to shape the education system in the post-Katrina era.[41] Implementing the reform should be regarded as an evolutionary process, instead of a revolutionary one, that will gradually create a supportive environment for collaborative decision-making, reduce pressure to reform, and generate positive effects of administrative interventions introduced in the educational system. Teachers and administrators need to accord much greater respect to students and their families, rather than treating them as unwelcome clients. These could be achieved through meetings and routine engagements with community leaders rather than just delivering resources immediately.

Competency

In rebuilding the U.S. Gulf Coast post-Katrina, little attention has been paid to the restoration effort in rebuilding the Gulf arts and cultural sector. The Gulf is marked by its rich cultural legacy, mostly internationally recognized for its music, literature, cuisine, and dynamic heritage. Its cultural legacy in the arts and culture reflects its ethnically and linguistically diverse population.[42]

As a $5 billion tourism industry, and the host of thousands of residents earning their livings as writers, musicians, visual artists, proprietors of artistic venues, filmmakers, architects, and curators of museums and historic sites, the Gulf cultural sector is critical to the region's overall economics and of moral value for its residents as a whole. In the wake of Katrina, many of the cultural workers in the region were displaced throughout the United States, which threatens vital social and communal networks that sustain the health and success of creative persons and communities.

The loss of cultural and economic vitality imposed risks and barriers to enhance the capacity of cultural workers and communities to function effectively within the context of culturally integrated patterns of human behavior defined by the group.

Cultural competence as defined by Georgetown University's National Center for Cultural Competence

> is a set of congruent behaviors, attitudes, and policies that come together in a system, agency, or among professionals and enable that system, agency or those professionals to work effectively in cross-cultural situations.[43]

The African-American psychologist Wade Noble defined cultural competence as follows:

> Culture represents the vast structure of behaviors, ideas, attitudes, values, habits, beliefs, customs, language, rituals, ceremonies and practices peculiar to a particular group of people. Culture provides ... (1) a general design for living and (2) patterns for interpreting reality.[44]

Louisiana has been deeply influenced by a mix of its French, Spanish, Native American, English, German, Nova Scotia Acadian (Cajun), and African ancestry. In the case of New Orleans, the storm affected distributed goods that are linked to cultural attributes. Pre-Hurricane Katrina, about 67 percent of New Orleans's residents were Afro-Americans.[45] Consequently, New Orleans's Afro-American majority has long been the core of its cultural and economic vitality; many Afro-Americans and poor residents, returned or otherwise, shared the cultural meaning of its goods—music, food, language, history, and art.[46]

Viewed in this way, the very fabric of communities' cultures as well as many of the racial and ethnic cultural activities practiced by families and individuals in New Orleans was cut down. Culture competence promotes a protective system that is comfortable and reassuring, especially in the face of adversity. For example, stories, rituals, and legends as well as cultural venues such as libraries, parks, schools, public media, and safe street corners, support a shared knowledge and information for people and communities to adjust to as survivors react to and recover from disaster within the context of their individual racial and ethnic beliefs, cultural viewpoints, life experiences, and values. Thus, culture competence plays an important role in disaster response since services are most effective when they meet survival's cultural beliefs and are consistent with their needs.[47]

Thus, public administration needs to incorporate culture into other public policy areas, such as education, health, and social services, so that Hurricane Katrina will not become the worst *cultural* catastrophe America has ever experienced dependent on the decisions made by those who administer the public trust.

The effort to recover from Hurricane Katrina seems to have spurred the growth of community competencies in New Orleans. Studies have shown that communities with greater collective competencies or social capital—cohesive communities, strong organizations, enthusiasm and mobilization, mutual trust—recover more effectively than those without collective competency.[48] Following Weil, people and solidaristic communities have appropriate competencies as "better educated and higher income people are more engaged, as are Jews, church members, and members of Social Aid and Pleasure Clubs (SAPCs)."[49]

Despite lack of material competencies, the Vietnamese community, united by the Mary Queen of Vietnam (MQVN) Catholic Church and Community Development Corporation, had already begun planning prior to the storm. The critical role played by the MQVN Catholic Church in community planning and recovery from Katrina fosters social cooperation and community rebound in the wake of disaster. Utilizing of unique bundle of community goods, residents in the New Orleans East Vietnamese-American community were able to preserve their distinct ethnic-religious-language community, triumph over coordination difficulties created by the storm, and become involved in successful political action to protect their community.

Pre-Katrina, MQVN's efforts were concentrated in Vietnamese language religious services, Vietnamese language education, and occasional weekend markets for selling Vietnamese produce, arts, and crafts that allowed members to establish a distinguished ethnic-religious-language community.

In the wake of Katrina, MQVN's efforts included building a retirement home in a parklike setting, accompanied by an urban farm and farmers' market. The community even convinced FEMA to build a temporary trailer park on the site, laying all the plumbing and electrical work in such a way that it could then be transformed into the foundation of a retirement center. The coordinating competency of the church was reflected by the high degree of overlap between leadership within the church and among secular civic organizations, including the Boards of Directors of the National Alliance of Vietnamese American Service Agency (NAVASA), Vietnamese Initiatives in Economic Training (VIET), the Community Development Corporation (CDC), and the Vietnamese-American Youth Leaders Association (VAYLA).[50] Through these collaboration initiatives, the church provided space for after-school tutoring, English language courses, Vietnamese language classes, youth leadership development, and business development. During the storm, the church's provision of civic space has become the most vital and common physical space, which facilitated the enhancement of social networks and collaborative efforts.

One of the admirable efforts presented by the Vietnamese community was the building skills of some of the community members who went house to house in teams, putting on new roofs, so that the residents could sleep dry in their houses, even while they worked on them. In addition to such building supplies, members helped in cooking and preparing communal meals for community members. Thus, within about six months in the wake of Katrina, most community members had returned and had a safe environment for living as a result of community's own competencies.

The Jewish Federation of Greater New Orleans (JFGNO) also engaged in extensive community recovery planning, building on a long-standing tradition of community competency. The JFGNO conducted a recovery survey in 2006 to assess community needs and interests and guide allocations. In addition, the Jewish community initiated a successful "newcomers" program to attract young, dynamic new community members to relocate to New Orleans. Due to its economic competency, the community was able to offer financial and communal incentives with event invitations to attract young members in both the business and the nonprofit realms.

The Latino community launched the Latino Health Outreach Project clinics after Katrina. The project began by setting up clinics on sidewalks and parking lots in areas where mostly Latino workers were gathering. It took no longer than a few months before the clinics were managed by MDs, nurse practitioners, acupuncturists, and herbalists. These clinics provide health care and empower community patients for access to health care across the city, supporting struggles for equity for immigrants and working people, and building collaborations with organizations that have a history of working in New Orleans's Latino community, as well as with poststorm response activities to ensure residents' right of return.[51]

When exploring the form of distribution, especially of material and social basic needs that aimed at securing the well-being of New Orleans's residents during and after Hurricane Katrina, scholars and practitioners considered nonprofit and grassroots organizations' support and care as a source of competency and empowering intervention. In the domain of justice, a grassroots coalition of organizations seeking to reform New Orleans's criminal justice system to build safe communities, met with several thousand hurricane survivors who were imprisoned at the time of the storm to establish a coherent strategic plan for safe streets and strong communities for everyone, regardless of race or economic status.[52] The Safe Streets/Strong Communities coalition provided legal support, and grassroots organizing that involves working directly with the imprisoned, former prisoners, and their family members. In its first months, the group had already succeeded in radically transforming the city's indigent defense board to an agency staffed with criminal justice reform advocates.

However, in too many cases, officials failed to coordinate with nonprofit community-based organizations and activist groups, waving red tape and rule books.[53] Such failure can be understood to be a result of lack of professional expertise, corruption, and lack of networking skills and clarity of the roles of private and nongovernmental organizations needed to make sound judgments in handling disaster policies.[54] In the case of Katrina, with FEMA's failure to satisfy needs for medical care, food, and housing, nonprofit organizations promptly stepped in to coordinate the response effort and met these needs by supplying immediate material infrastructure across the nation as well as locally. In contrast to administrative agencies, some of the nonprofit community-based agencies used a network of embedded ties to increase successful emergency relief efforts, such as the New Orleans Coalition on Open Governance (NOCOG), Rapid Evaluation and Action for Community Health in Louisiana (REACH-LA), and Mayday New Orleans.

It is suggested then that public administrators can be a source of a supportive environment that empowers individuals and communities actively engaged in community-administration collaboration, the same as volunteers and charitable individuals who became the immediate source of generating associative obligations and empowering intervention for the victims of natural disasters. The Walzerian principle of need is meant to show that equality of membership has a fundamental influence on the processes by which people control their lives and in the working out of the appropriate modes of provision. Voluntary associations of civil society, in which people share the view of need, are beneficial for community members who experience levels of necessity and stress. Thus, public administration should collaborate with local self-help and voluntary organizations in welfare provision to facilitate community members gaining power over the distribution of needed goods.[55] Without cultivating a relationship with a community through the participation of nonprofit organizations in decision making, city administration is unlikely to find active and interested citizens.

Disaster Resilience Management in Indonesia: The West Sumatra Earthquakes, 2004

The Civil Service of Indonesia

The civil service in Indonesia should be traced back to the roots of the Javanese administrative practices that regulated public affairs and established norms in a succession of the island's notable empires, including Sailendra Mataram, Majapahit, the Demak confederacy, and seventeenth-century Mataram. The Javanese administration was highly centralized and authoritarian based on a hierarchical structure of administrative chain-of-command by the master (*gustiand*, local ruler) and his servants (*kawula*).[56] This tradition lasted for nearly a millennium before submitting first to the European colonial practices, and subsequently to global standardization. Today, these administrative practices still bear some influence on Indonesian public administration. Traditional Javanese norms, such as centralized and patrimonial power, and top-down decision-making system and conflict solutions are entrenched in modified form as dominant features of Indonesia's public administration. Thus, the Indonesian public administration structure is formally hierarchical, composed of informal patron-client relations between public officials and citizens. The initiatives set down by the New Order aimed to limit the huge increase in civil officials, whose numbers stood at 2.5 million in 1968.[57] The New Order addressed regulations directed at tightening the central control over the bureaucracy, such as daily operations and clarifying the authority of central government over personnel of the regional government.[58]

The New Order followed the *Pancasila*, which was set by the 1945 constitution as the embodiment of basic principles of an independent Indonesian state. The *Pancasila* included five principles, among them belief in one supreme God, humanitarianism, nationalism expressed in the unity of Indonesia, consultative democracy, and social justice. These principles were articulated by Sukarno in a speech known as "The Birth of the *Pancasila*," which he gave to the Independence Preparatory Committee on June 1, 1945. The spirit of the *Pancasila* was also applied in the New Order reforms of the civil service by developing merit-based criteria to improve government capacity for economic development. For example, in 1978, a national indoctrination program was initiated to incorporate Pancasila ethics by civil servants. Suharto's New Order seems to pursue a mix of centralized and decentralized arrangements and to link them in ways that most effectively promote stability and economic development.

Currently, Indonesia's civil service stands at about 4.6 million employees. Among them, about 500,000 are employed in police and military agencies, leaving some 4 million civilian civil service employees. It should be noted that there is no formal job classification in the civil service. Entry ranks are mainly determined by education level, and increases in rank are largely driven by seniority—with a maximum rank depending on the entry level of the civil servant; significant discretionary allowances are distributed by top management in individual agencies to their subordinates in exchange for loyalty and, frequently, collusion in malfeasance.[59]

The administrations in China, Japan, Indonesia, and the United States have all put in place economic policies designed to increase economic growth and competitiveness, reduce poverty, and improve effective governance. In each case, these efforts responded to their own histories. Despite differences rooted in cultural, political, and historical experiences, the historical emergence of merit-based civil service was common in all cases. Among the four administrative systems, exams are administrated through the personnel departments and continue to be career-based, as their achievement is necessary either to remain in the civil service (basic exam) or to be promoted in a higher-level civil service. These administrative systems seem to have legacies of strong bureaucracies except in the United States. Until relatively recently, central government administrations played powerful roles in China and Japan, and still continue to be very powerful in Indonesia in the country's economic development. Such involvement in economic development was coupled with increased demands for public control and slow emergence of democracy in Japan and Indonesia, and to a lesser extent in China. Japan has experienced calls for democracy since the early 1970s, and China since about 2000. Given these developments, strong, autocratic bureaucracies of the past are no longer legitimized. Despite the preceding trends of democratization and transparency shared by these countries, their different histories seem to have a hold on their contemporary public administrations—and are very different from the U.S. public administration, which has roots in individual rights, freedoms, and persistent public distrust of central government. At the same time, democratization and distrust of strong central government could bridge the cultural and historical gap between the United States and Asian public administrations. Thus, public administration would need to establish better communication channels for extensive deliberation and consensus building between citizens and the government for citizens to have an effective voice in all stages of the public-policy process.

Advocacy

The aftermath of the tsunami in Aceh province and Nias Island, Indonesia, on December 26, 2004 destroyed hundreds of thousands of buildings, leaving approximately 190,000 homeless, and 67,000 people (including children) living in barracks or tents.[60] The 2004 disaster provided an opportunity to understand how community stakeholders—the target for postdisaster settlement and shelter advocacy—interpreted and gave a sense of worth, power, and ownership to those needing transitional settlement and shelter. Such understanding is important, as advocacy may be considered an effective means of both image building and influencing policy, because it is able to persuade without seeming to do so. As a means to raise awareness by appealing to commonly shared values, it promotes community involvement in decision making. In post-tsunami Aceh, advocacy campaigns were initiated to communicate communities' stance on implementing settlements and shelter in an effort to enhance self-help initiatives and their contribution to the health and preservation of the community while indirectly reducing the potential for government intervention in corporate construction activities.

Through communities' active involvement in the reconstruction process, pressures arose from expectations of communities and of *Badan Rehabilitasi dan Rekonstruksi* (the institution in charge of coordinating Aceh's reconstruction) to use Western-modern style rather than timber dwellings, as they symbolized a more developed and progressive image even at the cost of safety and security. Indeed, in Aceh, timber has been the traditional and self-building method, and proved itself as an effective and safer construction technique than self-built masonry houses because timber techniques were more practiced and rested on existing knowledge and skills.[61] Despite the fact that the modern structures were less suited to the region's climate and traditional knowledge about settlements and shelter, communities in Aceh involved in reconstruction advocacy referred to the economic and emotional benefits they may potentially gain from it, such as employment and community and regional pride. For example, in Aceh, Besar residents have addressed the role of external appearance of masonry houses, which represents modernity and the affluence associated with middle class.

Since advocacy is by definition value laden and subjective, it led communities to use subjective viewpoints regarding the meaning of settlements and shelter in choosing between masonry or timber building methods—such as aesthetic or modern, rather than safe. However, the involvement of community in implementing settlement and shelter resulted in greater difficulties in meeting the requirements of security with livelihoods, especially fully incorporating environmental considerations where houses are conceived as an integral part of safety, security, livelihoods, and of other sectors, including water and sanitation. Findings of a survey conducted in post-tsunami Aceh from 2005 until mid-2007 point to the fact that the task of balancing these various concerns and trade-offs was frequently forgotten due to community involvement in the reconstruction process and its unrealistic expectations of modern reconstruction.[62] Despite commitment to provide large compensation grants and support by international NGOs, Aceh suffered from a lack of professional or experienced construction staff who did not follow the international guidelines for transitional settlement and shelter to ensure sound technical advice for safer rebuilding.[63] This could be partly explained by the fact that Aceh was isolated and was only engaged in a few international NGOs implementing projects. This problem was further intensified by the fact that many government offices were destroyed in the disaster, which resulted in poor coordination among organizations and lack of coherent and consistent reconstruction policies. The lack of experience of both the Acehnese and Indonesian governments of such a large-scale emergency paved the way for ad hoc and rapidly changing policies. In addition, the implementation of transitional settlement and shelter encountered oversight of environmental impact assessments and long-term spatial, urban, or regional planning, as well as political and security issues contributing to a high likelihood of growing vulnerabilities, which were exposed by the tsunami.[64]

Thus, community advocacy should be enhanced through proper information and adequate communication to ensure that tsunami-affected communities' expectations are considered and balanced with issues of safety, security, and livelihood to reduce frustration and tension that resulted from unmet expectations. It is then suggested that community advocacy should involve active participation of community representatives in decision making for settlement and shelter, rather than overall control of the decision making. Moreover, organizations engaged in transitional settlement and shelter building efforts should consider local capacities of the target area, other sectors, and environmental considerations involved in order to reduce long-term impacts on ecosystems and livelihoods.

A more successful example of community advocacy was realized in the domain of reconstruction of displaced communities in post-tsunami Aceh. In the aftermath of the December 2004 tsunami, the Indonesian government developed a relocation solution program called "barracks camps," also known as "temporary relocation shelters." The Indonesian government's plan was to relocate about 60,000 tsunami victims into thirty-six supervised barracks.[65]

Although international organizations fiercely supported use of the barracks on humanitarian grounds, the issue of barracks (*baraks*) has been deeply controversial. As a contested program of humanitarian intervention in post-tsunami Aceh, barracks camps provide an enlightening example of how a particular solution to "the problem of displacement" was discredited in the eyes of victims.[66] Viewed in this way, the policy solution of the barracks camps forced affected communities to better articulate and evaluate policy options based on their shared values and interests. These camps served as networks between victims for building relationships and collaborations among these "internally displaced persons" who share specific rights and needs under international humanitarian and human rights law. Victims' engagement in such barracks helped them to mobilize their collective needs and press the government to resolve their housing problem. They have succeeded in mobilizing collective protest against government failure to supply novel opportunities for greater social justice and in affirming new societies forged in the context of displacement in post-tsunami Aceh. Throughout 2006, demonstrations that involved thousands of people called to challenge government regulations and the deconstructing reconstruction in post-Tsunami Aceh resettlement of barracks IDPs.[67] Despite the fact that in 2006 tens of thousands displaced by the tsunami still dwelled in barracks, the Indonesian government began to work on programs to address the needs raised by the protesters. The emergence of barracks as sites of advocacy laid the ground for understandings of the very nature of the "problems" of displacement and the displaced.

Inclusion

The impact of the 2004 tsunami was not equally distributed over the population of Aceh and Nias since some groups of people are more vulnerable than others. Hence, the most vulnerable groups affected by the negative impacts of the disaster damages have the least capacity to resist and recover from the losses imposed by the disaster. Among the most vulnerable groups are women and children, persons with disabilities, and other socially excluded people. In both Aceh and Nias, remote communities with minimal service provision predisaster, children and women, especially poor female-headed households and war widows, and communities were affected by violent conflict. Although post-tsunami recovery programs addressed these groups by providing them temporary shelters, disaster emergency response programs did not provide adequate attention and very often they were systematically disadvantaged for various reasons (cultural, economic, etc.).[68]

The process of exclusion of these groups in the Indonesian disaster management setting kept them away from active participation in the wider economic, political, cultural, and social life domains; led to discriminatory practices on the basis of their ethnicity, religion, gender, age, disability, and migrant status; and limited access to decision making that could influence policies or create opportunities for improving their standard of living. For example, the remote and underdeveloped coastal zones of Aceh and Nias, deprived of basic infrastructure and the dispersed nature of local settlements, made marginal fishing communities in Pidi, Aceh, Jaya, and Nias vulnerable, as they did not receive any assistance. For the Acehnese in general, and the people of Nias in particular, development and relief efforts lagged far behind development elsewhere in Indonesia. People in these regions felt marginalized and isolated from the national overall development process, economically as well as politically—especially in Aceh, due to its long sociopolitical conflict, and in Nias, because of the separation from Sumatra and the geographic remoteness of the island.[69] The province of Aceh experienced both natural disaster and man-made disaster. For decades, Aceh experienced civil wars, insurgencies, military operations, and separatism that resulted in high numbers of casualties, disability, and long-lasting mental scars.[70] Through these armed separatist conflicts, the government of Indonesia and the Aceh Independence Movement (GAM) were not the only victims of that conflict, but also their families, siblings, and children who bear directly or indirectly the negative impact of this violence. The natural disaster thus intensified the region's instability felt by both sides and by the families of these people.

Since the 2004 tsunami in Aceh, children faced long, negative, psychosocial stresses resulting from losses of family members, homes, relatives, and good friends, and stayed in temporary barracks and tents with a low standard of living. In addition, they had to adapt to various changes such as new neighbors and friends from different regions, new schools, new teachers, and family and community structures.

In an exploratory study that employed observations and interviews involving 117 children aged eleven to fifteen years old, from three camps in Lambaro Skep, Jantho, and Tanjong, located in Banda Aceh and Greater Aceh areas, children have shown greater adaptation to changing realities. Local intervention programs such as routine religious activities seemed to help children cope with their trauma.[71] The research findings have shown that participation in community activities made them feel cheerful, able to help each other, actively involved, able to speak confidently in front of people, and show other positive behaviors such as responsibility and care.[72]

In the face of adversities faced by children, the inclusion of children was also enhanced by CBDP initiatives to involve children in preparing the Child Survival Kits and help their mothers to get the Family Survival Kits ready. Through this activity, children gain the PLA exercise and know about the safe and vulnerable places. They were also trained in simple first-aid skills, including preparing ORS. *Makkala Panchayat* (children's *panchayat*) was formed as a village-centered program to address the problems faced by children, in collaboration with the local government and with the participation of children. Such inclusionist initiatives provide a proper arena for children to share their anxieties, needs, and skills in overcoming difficulties within significant adversities after trauma and also to reduce the work burden and anxieties of women.

Studies on the 2004 tsunami reveal that the women and children became more severely affected than men because of their daily activities, which confined them to homes, and are less mobile than the males, which increases their vulnerabilities and sufferings during such incidents. Aside from physical and psychological traumas caused by the fact that families were broken up and moved to IDP camps, women were often vulnerable to sexual harassment before and after the 2004 tsunami.[73] Since women concentrate their attention and relief efforts on family members, they have tended to accord much lower priority to disaster risk management activities in the presence of more pressing community concerns.

In the area of cultural exclusion, post-tsunami Aceh religious authorities, especially the shari'a, have been enforcing regulations more strictly. These stringent restrictions focused on women's behavior in the face of adversity. In the area of economic security, female survivors experience difficulties with land titles, inheritance, guardianship, low economic base, and absence of capacity to raise funds. The awareness for women's inclusion in participatory decision-making to reduce these vulnerabilities was raised by the Komnas Perempuan (KP), the National Commission to Prevent Violence against Women. The KP appointed a special coordinator to monitor the response efforts in Aceh, focusing on settling land disputes and challenging the shari'a strict regulations.[74] KP proposed a new procedural law on land rights addressing the local wisdom as articulated by the shari'a canons and promoting open dialogue with the judiciary and government on its interpretation—for example, on the need for separation of men and women. The KP joined other local female and human rights NGOs such as the *Purta Kande* and the Indonesian Women's Association for Justice (APIK) to collect and disseminate information regarding forms of violence against women, provide moderate advice to the formulation of the canons (local legal regulations that acknowledge Adat and Shari'a laws), and lobby officials.

There have been other women's inclusion initiatives worth mentioning, such as the EAD and Flower Aceh, which have provided microcredit funds, economic business activities, and marketing projects for vulnerable women's groups affected by both conflict and tsunami. The United Nations Development Fund for Women (UNIFEM), the United Nations Economic and Social Commission for Asia and the Pacific (UNESCAP), and UNICEF, among others, work on gender-segregated data to address the needs of most vulnerable women. The UNIFEM program focused on the following four areas to support women's inclusion:

1. establishing credit initiatives and creating places to work for women living in barracks;
2. protecting women from religious and cultural sanctions by setting up a working group in collaboration with UNFPA, as imposed by the shari'a law security guidelines for legal aid clinics and camp staff and police;
3. targeting women's groups overlooked by the international tsunami response, such as the affected war widows in Pidie district; and
4. promoting women's leadership through the appointment of gender advisers to work with BRR, to consult with government and INGOs to guarantee women's participation, to map women's CBOs, and to collaborate with the less powerful Women's Empowerment Bureau.

The awareness raised by these initiatives advanced the inclusion of women in disaster management so that gender needs are taken into consideration (significant representation of women in decision-making bodies, leadership training for women, and building strong women's collectives). Gender and disaster training would assist policy makers to understand why a gender perspective is needed in the policy-making process and how gender-oriented interventions can be incorporated in any program in the context of both development and disasters. Thus, by gaining nontraditional skills women will be able to transform the perceptions of the community toward women and bring new perspectives and solutions in facing future adversity.

Competency

Community competencies during the 2004 tsunami disaster were evident even when people were still living in temporary communities or camps. Identifying various competencies that can improve community members' quality of life in the face of adversity will also provide a solid ground for the rebuilding stage to follow.[75]

An important contribution that was made to building more protection and resilience in the face of future disasters is the establishment of a community warning system. In Indonesia, community leaders used the *Smong* ("Inamura-no-hi"), local verbal tales that inform listeners of the danger of disaster as a tool to raise public awareness. Such a community-based device was proven efficient in saving the Simeuleu communities from the Great Indian Ocean Tsunami in December 2004. In Banda Aceh, people had no idea about the risk and even summoned relatives to help "collect fish" from the beach, when the sea suddenly drew back after the December 26, 2004 earthquake. On Simeulue, the islanders ran immediately to the mountains instead, leaving everything just as it was. Only nine people died on the entire island.[76] This community initiative is considered nimble and flexible, and it can reach out and protect at-risk communities almost immediately to the extent that they are able.

Community competency refers to interpersonal and group linkages, social support, and community bonds. Interpersonal and group connectedness are viewed as a common social memory of the community to remember experiences from disasters,[77] which creates fertile ground for innovative learning and better adaptation to future adversities.[78] Social support connotes the social interactions of individuals that in turn generate a social environment of mutual assistance, caring, and support. According to Aldrich,[79] in the aftermath of disasters people who are better connected to others receive more assistance than less connected people. Interpersonal connections that enhance better communication between people and allow individuals to share information about community needs and resources involved in disaster assistance procedures were identified by villages and towns that initiated their own rescue and relief efforts in the spirit of the Acehnese tradition of *gotong royong* (voluntary mutual assistance).[80] The *gotong royong* handled numerous local projects organized around the value of community bonds and solidarity, including raising a roof, contributing to cultural or religious ceremony, or clearing a road. These measures enable affected community members to be emotionally connected to their communities, voice their concerns, alleviate potential limitations of population diversity, and collaborate to enhance equitable distribution of roles and responsibilities within the community. An admirable example of self-organization and self-help is featured in Krueng Sabee, a trading village on the sea whose community managed to reorganize in a short time by coordinating relief aid and planning and organizing recovery and reconstruction of their village.[81]

The tsunami killed 20 percent of the community in Krueng Sabee (1,480 villagers); nearly 70 percent of the victims were women. The 5,500 survivors were deprived of any assistance. On January 10, a group of villagers managed to alert aid helicopters and, at last, food drops were made. On January 12, two village leaders took the lead and arranged an effective system of food drops by negotiating with helicopter pilots on the times of food drops, so that supplies could be collected by the leaders and distributed each day according to village needs.

On January 13, the village leaders established a *posko* (community command post) to represent the people in daily coordination meetings with aid workers. The *posko* entailed collection of accurate information on surviving families and needs, and coordinated the search and rescue (burials) along with coordinated emergency efforts and food distribution. During February, schools were opened in tents. Villagers' activities included making furniture from waste materials and building a local mosque, which was used as a gathering place for community meetings and information exchange. In addition, the *posko* promoted building partnership and collaborations with INGOs, government, and local leaders to plan how to best rebuild and protect the villages from future adversities. The community competence of Krueng Sabee outlined four factors that contributed to its successful recovery from the 2004 tsunami:

> The culture of working together in the town was preserved from the first days after the tsunami; Strong village leaders survived the tsunami; Krueng Sabee is an historic centre of trade and entrepreneurship; and Data collection in the first weeks after the disaster gave the community a better perspective of its needs and provided a basis for proactive dialogue with humanitarian aid agencies.[82]

This case highlights the role of community leadership and community decision-making institutions that should be strengthened, and their capacity developed to manage the reconstruction and recovery process in their own areas. As seen, social and economic development initiatives were undertaken immediately through community collaboration. Community self-help efforts can be extremely effective both in improving the spirit and morale of communities, and in transforming the material conditions and improving the overall quality of life. Moreover, this case emphasizes community (village) leadership as a key factor that influenced the success of communities in rebuilding their lives in the face of natural disaster. Village heads in Aceh are called *Keucik*, or *Geucik*. The *Keucik* gains a prominent political and social role within Acehnese village society. The *Keucik* is elected and trusted by the community members and is officially appointed by the district/municipality government to manage the *Gampong* administration.[83] Aceh village leaders are assumed to be endowed with supernatural and spiritual powers (*kesaktian*), heredity (*keturunan*), knowledge (*ilmu*), and equity (*adil dan jujur*), and to be courageous and decisive (*berani dan tegas*), generous (*dermawan*), and kind and hospitable (*ramah tamah*).[84]

Historically, the *Keucik*'s decision-making power was restrained by a permanent council of elders (*cerdik pandai*), called the *Tuhapeut*. The *Tuhapeut* council functioned as a deliberative body independent of the *Keucik*, as the primary deliberative body in the village whose decisions on village affairs are reviewed and considered by the *Keucik*. It should be noted that the office of *Keucik* has gone through many changes over the past several decades. The armed conflict has had an impact on the role of the *Keucik*, which was often found to be less trusted and the target of suspicion by both sides of the conflict—the Indonesian military and police and GAM members. Under the 2004 tsunami, the *Keucik* and neighborhood heads (*Kadus*) who served as section heads (*Kaur*) under the *Keucik* were ill-equipped to meet the needs of their effected population—that is, if the officeholders survived the disaster.[85]

Economic competence is also a crucial capacity of community that allows community members and organizations to have job opportunities in fundamental needs such as roads, housing, accessible health-care, schools, physical capital, and machinery. Coastal communities are conceived as having more chance to diversify their economic actions in comparison to inland communities that rely on limited types of economic activities. Community-driven development—a strategy used by the World Bank and the Urban Poverty Project (UPP)—utilized projects that increase community power over development.[86] Increased community control on resource allocation in the Indonesian setting has led to the establishment of a new community-based organization (*Badan Keswadayaan Masyarakat*—BKM) or strengthening an existing organization to receive and administer the project funds. This arrangement of BKM organizations took about six months, depending on differences in factors such as a community's organizational capacity. In all communities, the BKM exists at the same level as the *Kelurahan*, the lowest level of the state's political administrative hierarchy. The BKM is organized by community volunteers who are selected by local community members through a deliberative democratic process with the aim of inclusion of poor residents. Each BKM has the power to decide, from three programmatic options, how to distribute its funds to diminish poverty: providing microcredit to local businesses; establishing physical infrastructure; and improving human resources through training. Most BKMs used the funds to invest in a revolving fund for microcredit for which groups of entrepreneurs develop loan proposals. The BKM functions as an independent from the state, thus utilizing a new political sphere.

Other projects that enhanced community economic competency were introduced by local government and NGOs. Provincial administration officials offered to distribute the state fund for reconstruction of damaged homes and office buildings directly into each community's bank account;

three months after the earthquake hit the West Sumatra coast no apparent reconstruction or rehabilitation programs by the government have been seen.[87] The provincial disaster management unit head advanced this program by requiring communities to list expected needed goods and services, including rebuilding infrastructure, funding social activities, and looking after survivors. Such administrative effort demonstrates empowered expression of genuine partnership between administration and community in welfare delivery in times of emergency.

In addition, Quake Fund joined the local NGO PADMA (*Pelayanan Advokasi untuk Keadilan dan Perdamaian* Indonesia/Advocacy Services for Justice and Reconciliation) to provide seventy families with secure and durable temporary shelters. PADMA Indonesia is a nationwide NGO that was established in 2002, and covers a wide range of issues within Indonesia. This project has followed PADMA's unique method of implementation that aims at strengthening local community capacities during and after the Aceh tsunami and the Nias earthquake. PADMA supports each family on a case-by-case basis, meeting their needs in terms of materials, labor, or both. PADMA also works to ensure that all participants achieve a safe secure shelter that meets or exceeds local and international standards, while utilizing as much recycled and locally procurable material as possible. Taking families and community members' needs into consideration, PADMA was able to work directly with the affected community, family by family, discussing their shelter needs and using this as an opportunity to discuss long-term plans for more disaster resistant, sustainable construction.

In contrast to these efforts to enhance community competencies in the face of adversity, some community initiatives were ignored and the knowledge, experience, and vision for recovery were overridden by the government, which was unable to rely on community self-help and sustainable recovery. During the tsunami aftermath a New Order in Indonesia was initiated by the state to replace local and traditional institutions including *Gampong* in Aceh, which was positioned under the official hierarchy of bureaucratic institutions. *Gampong* in the traditional Aceh system functioned as part of *mukim* territory, which under the New Order was reduced to being merely an administrative unit below the subdistrict level. Besides representing geographical territoriality, in the traditional history of Aceh, *Gampong* also meant a unit of "masyarakat adat society with territoriality."[88]

As these traditional functions of the *Gampong* had weakened, they could not use their power to prevent the development projects that entered remote areas in North Aceh and in the Gayo highlands leaving the *Keucik* and other *Gampong* members with no income from their *seunebok* or productive land, different from what had been the case in the past. Thus, the *Gampong* communities became poorer and their political participation had been weakened sooner.[89] Consequently, the social recognition of their *adat* role, which was crucial to maintain community competency and solidarity, eventually diminished.[90]

In conclusion, disaster response presents a huge opportunity to build human and community capacity for self-reliance and sustainable well-being and prosperity, if efforts of constructive engagement take into consideration communities' knowledge, experience, and vision for development and recovery needs.

For such initiatives to be effective in strengthening community self-reliance, they need to be guided by organizations that are community-driven and experienced in facilitating capacity building for self-reliant social and economic development, and implemented by organizations that are committed to enhance communities' capacity for collective action, such as the ability to cooperatively deliver public goods and services, and the capacity to redefine power relationships and create structural change when needed or recognized by community members.

Disaster Resilience Management in China: The Wenchuan Earthquake, 2008

The Civil Service of China

Service in the public sector carries with it considerable prestige in China. The civil service in China can be traced back to the Qin Dynasty (221–207 BC). In its early days, public administration in China, especially the military, was based on merit. Officials were evaluated by their loyalty to the emperor, and their understanding of Confucian philosophy and practice.[91] China's ancient merit system, namely the *Li*, was established by the civil service examination system that equipped them with administrative competence of the *Li*.

Since the establishment of the People's Republic of China (PRC) in 1949, the administration served the Chinese Communist Party (CCP), which had direct control of administration. The administrative system was driven by the communist ideology and focused its efforts on class struggle and national security.

The public administration after 1978 is associated with increased modernization, such as specialization, meritocracy, due process law, and accountability. In 1988 the Ministry of Personnel was established, and a year later the central government announced state civil servant examinations to replace the old labor allocation recruitment system of lower-rank administrative positions.[92] In 1993, the provisional Regulations for State Civil Servants were circulated to replace the cadre system. The transition from cadre management, regarded as a process of administrative downsizing and recruitment, was based on open competition, usually through an examination process and limited for the most part to university graduates and qualified staff from the public sector. Staff development reform focuses on performance appraisal and training along with career structure and stable employment. The State Administration School was founded for both administrative research and education in 1994. Many civil service training programs involved overseas excursions and generous spending, while leading to growing suspicions and public concern over the efficiency and effectiveness of these training programs. These reforms finally led to the adoption of the State Civil Service Law in 2005, with clear emphasis on professionalism. Political neutrality is not addressed by law, which leaves the impact of party leadership on personnel policies and management.

In comparison to public services overseas, the Chinese definition is both more inclusive and less inclusive than definitions of the civil service commonly used in other countries. The public service in China refers to public employees in people's governments, people's congresses, people's political consultative conferences, courts, and procuratorates of various levels. The Chinese civil service stands on over 70 million people, half of whom worked in government in one capacity or another, with 33 million working in state-owned enterprises (since 2003).[93] Thus, contrary to Western countries, the civil service in China includes the most senior politicians such as the premier, vice premier, state councilors, ministers, and provincial governors, vice ministers, and vice governors—the leadership positions.[94] At the same time, the scope of the Chinese civil service is exclusionist; it excludes all manual workers employed by the government, and the employees of all public service institutions, including schools, universities, hospitals, research institutes, radio and TV stations, cultural organizations, publishers, and so on.[95]

Advocacy

The magnitude 8.0 earthquake in China, considered one of the country's worst natural disasters in more than thirty years, has temporarily altered the world's perception of China. The media coverage reflected a more compassionate and generous nation than many perhaps expected, where nations praised China for the speed and scale with which the government responded through the relief operation. However, this new self-awareness of the Chinese society cannot easily erase the fact that China's economic and military growth was at the cost of slave labor and child-kidnapping rings, rampant government corruption, counterfeit products, tainted food, dangerous toys, and, lately, the brutal attack on dissent in Tibet.

In theory, advocacy can emerge if public officials learn how problems of individual interest and community welfare may both be effectively addressed through government relief programs. In day-to-day engagement, practitioners should struggle to create real deliberative spaces in the contentious and political world of planning so that notions of interest and community will not become politically shaped—not only by public officials' imaginations, but by those who do not speak.

In China's autocratic regime, communities' interests and concerns generally have limited access to political, legal, financial, or scientific resources. Communities do not develop the necessary resources to enable protest or even movement formation, and this lack of resources is especially obvious when compared to the well-documented ability of communities to influence public opinion or government agencies in China.[96] The existing governmental frame of ever-expanding growth and development is hegemonic to the extent it is taken for granted as "good," normal, and tolerable by the Chinese people. Thus, hegemonic frames are hard to challenge or protest. A community protesting disaster management, fighting the specific proposed policy, is also fighting general societal values and norms. It requires a powerful community and civil society to alter the great power that lies in the hands of government agencies to achieve regulatory frameworks more beneficial to the local community.

During the 2008 Sichuan earthquake, allegations of corruption in the construction of Chinese schools were raised by grieving parents and critical journalists. According to official estimation, more than 9,000 schoolchildren and teachers were killed by the Sichuan earthquake. They accounted for 12 percent of the total number of victims. Over 7,000 poorly built schoolrooms collapsed and another 14,000 were damaged.

Parents gathered day and night for almost two weeks around the collapsed school buildings waiting to learn the fate of their children.[97] On May 29, 2008, government officials began inspecting the ruins of ten collapsed schools in the city of Shifang. In other places, investigations were not far-reaching, implying that the case was closed; often investigations were not carried out at all.

In September 2008, the Chinese government admitted that pressure to meet the growing demand for schools may have led to rushing implementation of building programs at the cost of a lack of reinforcements in columns that supported classrooms, poor urban planning, and less strict building standards for schools. The government defended its shortcomings by exaggerating the intensity of the earthquake. The public criticism led to use of the term *tofu buildings* to denote the poor quality of these constructions that caused immense death and suffering for innocent schoolchildren and practitioners and their families. Despite the immediate intention of the Chinese government to sweep the school scandal under the rug, much anger and dissent was directed at the government's failure.[98] Human rights groups accused the government of concentrating on silencing anger rather than carrying out thorough investigations and blaming those responsible for the poorly engineered construction.[99]

For example, Huang Qi is a human rights advocate who served as a voice for the families who lost their children during the Sichuan earthquake by publishing numerous articles about the shortcomings of the schools' planning and building. In June 2008, Huang Qi, as well as Pu Fei, a volunteer for Tianwang, and Zuo Xiaohuan, a former teacher at Leshan Teachers College, were detained on charges of "illegal possession of state secrets."[100]

On July 25, 2008, a teacher at Guanghan Middle School, Deyang City, Sichuan Province, who traveled to the affected areas in Sichuan, took photographs of collapsed schools and posted them on the Internet, was detained on charges of disseminating rumors and violating the social order and sentenced to a year of reeducation in a labor camp. His family turned to international human rights organizations to help gain his release after he was being denied family visitation.[101]

Tan Zuoren, an activist who investigated the deaths of thousands of children who died in the quake, was sentenced to three years in prison on unrelated charges. Zeng Hongling, a retired teacher, was detained after publishing essays criticizing the government's earthquake response while putting the blame for the school scandal on official government corruption.[102] The artist Ai Weiwei has used his blog on Sina.com to post criticism of the government's response to the Sichuan earthquake. Although the government put strict restrictions on such reporting of the school scandal, foreign journalists tried to interview parents to enhance their advocacy to raise their voices. In most cases, these journalists found themselves detained for "working behind police cordons"—interfering with the government relief efforts.[103]

Parents of Sichuan earthquake school victims joined together to meet with officials from the ministries of education and construction. The government representatives in Beijing made a show of listening to their concerns, but concentrated their efforts to silence the angry parents through financial compensation in exchange for their dropping their demands.[104]

Admirable advocacy initiatives were taken by parents whose children died during the Sichuan earthquake in Dujiangyan; 100 parents protested the deaths of their children in poorly built schools. These parents protested in the name of the 270 children from a high school who died in nearby Juyuan. The parents, marching to government headquarters carrying pictures of their dead children, called for filing suit against those who were responsible for their children's deaths. In Hanwang, parents were detained by police for twelve hours after attending a protest. Parents who refused to sign the compensation contract were detained for longer periods. Advocacy grew further among the grieving parents who tried to sue or petition local and central authorities together with human rights and activist groups and lawyers that have tried to help them. In December 2008, a lawsuit demanding an apology and compensation from the school system and local authorities was filed by parents of 58 children who died in a collapsed school was dismissed by a Sichuan court. Parents from Fuxin filed a lawsuit against government officials and a construction contractor, demanding $1.1 million for physical damages and a stated public apology. A judge at the Intermediate People's Court in the city of Deyang dismissed the lawsuit by following government directives. In addition, a group of parents from Mianzhu secretly traveled to Beijing and filed a petition at the central government's petitioning center three times.

As parents' advocacy grew more persistent with parents' protests, the government became more aggressive in its efforts to suppress any dissent or criticism of negligence or corruption on the part of government officials. The suppression of parents' protests led the government to offer monetary contracts. The financial compensation amounts vary by school but are roughly the same. For example, in Hanwang, parents were offered a package valued at US$8,800 in cash plus a $5,600 pension in exchange for their silence. Similar offers were made to parents of students who died in other collapsed schools. Many parents were forced to sign monetary contracts, even if no real investigation was launched. Furthermore, officials have used traditional methods of silencing such as harassing, detaining, and physically threatening to break up demonstrations of parents who did not accept compensation for the death of their children in the Sichuan earthquake.[105] Daily, the local government watched the school site and roads leading into the village to report on suspected parents and impede local and foreign reporting on school collapse or raising the issue again on the public and international agenda.

As seen, the Chinese government has expressed a zero-tolerance attitude toward public criticism over the school collapse during the Sichuan earthquake. The school scandal in China has proven that growing distrust of government promises and claims of concern for the community or public interest can lead to feelings of uncertainty, suspicion, and dissent on the part of the citizens. Although the direct voice from the community—especially from the groups of parents whose children died during the 2008 earthquake because of poor school construction—has often been disregarded by the government, it is a pivotal component in advancing community advocacy in China. Parents' groups were able to translate their pain and anger (as well as expectations) into sporadic collective actions that speak to the economic and political concerns of the government. As a response, government agencies viewed foreign grassroots activists and parents' groups that complain about the responsibility, corruption, and negligence as irrational or hysterical.[106] These complaints were framed by the governments involved as "perceived" and unproven claims against "real" and proven disaster intensity and immense damages. Community perception of school collapse was strongly correlated with impressions of vulnerability emerging from distrust of both government and administration. Parents' group concerns become a type of social problem, calling into question the autocratic structure of government that limits citizen input, participation, and control. Despite a successful campaign to silence parents and social activists, local community residents still hold suspicions of both government and administration, convinced that neither has their best interests at heart.[107]

Inclusion

The 2008 Sichuan earthquake had a unique impact on powerless groups of people, such as older people, children, women, and ethnic minorities, reflecting the socioeconomic environment in rural China, where many people of working age have migrated to urban settings for better job opportunities, leaving behind a population in which older people, women, immigrants, and children are overrepresented. In the immediate aftermath, the government provided free access to services, irrespective of place of origin or status. Yet migrant workers, who officially make up 15 percent of the population in the affected area, were more adversely affected by mental and physical vulnerability due to lack of entitlements to free medical and socioeconomic services.[108]

Women appeared to be more vulnerable with respect to the psychological effects of the earthquake.[109] Studies show that natural disaster reinforces the powerless of women who are already in a vulnerable position within the family in their ability to cope with trauma sequelae when exposed to a disaster. Researchers addressed the factors of lack of social support and resources, as well as lower economic status and lower education as the factors most associated with the vulnerability of women exposed to trauma.[110] Women who are less educated and less skillful are found to be less resilient in recovering from trauma because of their poorer capacities and skills, lower self-esteem, and lower levels of insight.[111] In addition to these effects, intimate partner violence (IPV) rates increased during and after the disaster, as shown by a recent study conducted by Anastario et al.[112]

Natural disasters thus affect women's vulnerability, due to their inferior financial role, gender role stereotypes, and cultural norms and perspectives.[113] It is claimed that stress may affect the general state of deprivation, increase sensitivity to negative stimuli, and lead to more frequent family violence incidents, which finally contribute to criminal behavior, such as aggression toward others.[114] Therefore, stress resulting from adverse circumstances increases the incidents of disruption, conflict, and sensitivity to negative stimuli, and may lead to more frequent interactions with other depressed persons, which finally contribute to manifestation of criminal behavior such as aggression toward others from the close environment. In a survey conducted in June 2008,[115] in one of the temporary shelters in Sichuan province, the Du Jiang Yan Community A, findings have been in line with previous studies,[116] outlining that women who reported psychological victimization have consistently been found to be at greater risk for mental disorders, including depression, anxiety, substance abuse, shame, and fear, which imposed barriers in coping with disaster effects and quick recovery from trauma.

Successful community inclusion projects were exercised by the Qiang women and peasants' communities in Sichuan province. Pivotal to the project's success was the employment of a top-down structure in day-to-day life for people and communities with women, homeless, and peasants. The community of Qiang women in the Maoxian County of Sichuan was left behind as their husbands and sons have left the rural villages to become migrant workers in China's major coastal cities. The earthquake had devastating effects on these women, increasing their stress and loneliness, leaving Qiang women and single-mother families extremely vulnerable. In order to help the earthquake-affected Qiang women the Asian Foundation–China, together with other Sichuan Administration support, implemented community capacity-building activities including strengthening their mutual-assistance and networking capacity, and enhancing their participation in social life and long-term community development, to fully include this community of homeless and isolated Qiang women.[117] This case illustrates the principle of need-extended welfare administration to volunteers; the practice of active involvement in welfare engenders a sensitivity to need that is grounded in the "inner logic of communal provision" that is the principle of need.[118]

Another successful inclusionist project that also relied heavily on government initiatives was employed within the peasants' rural communities. During the postdisaster construction of new rural communities, peasants showed great enthusiasm and took an active part in reconstruction work.[119] The government used mechanisms of mobilizing the initiative of broad masses of peasants to become involved in the implementation of the sustainable postdisaster reconstruction as an effective way to accurately meet the real needs of local people.[120] These mechanisms included adopting a collaborative approach to addressing challenges and services gaps in rural reconstruction, promoting contact with rural elites to generate opportunities for inclusive participation, and developing an awareness of community activities at the individual, house, and province levels. Although peasants' communities were often excluded from decision making, to maintain their passive dependency to a certain extent, the performance of the government in rural reconstruction met with peasants' satisfaction.[121]

Rural elite played an active role in reconstruction efforts and collaboration with government officials due to the fact that they are more educated, skilled, and have close ties with both the masses and the government officials. Thus, lack of education and proper skills to gain a comprehensive sense of the overall situation is a major obstacle to peasants playing a genuine inclusive role in postdisaster reconstruction decision-making.

In sum, affected and socially marginalized communities have the power or at least the potential to create power through collaborations with NGOs, administrative officials, and other stakeholders. However, having power or the capacity to have power is still far from using power, so that the community must have the will to exercise their power through active involvement in creating economic, social, and political pressure to force changes more beneficial to the community.

In autocratic regimes, increased community inclusion raised uncertainty on the part of government agencies regarding including socially demanding issues that could lead to unwillingness on the part of the government to change existing regulations and standards. This can result in the unfair burden of disaster risks and vulnerabilities borne by a minority community. This section outlines the contradictory role and strategies regarding community inclusion in the face of adversity. On the one hand, government officials are expected to represent the best interests of affected communities. On the other hand, when state government officials and agencies ignore local and at-risk community complaints and concerns, these communities can turn to protest in order to have their voices heard in court and by the government, fueled and supported by foreign media and human rights advocacy coalitions.

Competency

The unprecedented wave of voluntarism, including volunteers and civic associations who joined forces during the earthquake relief, has swamped the affected areas in China.[122] China gracefully accepted aid, such as international NGOs working side by side with government-organized NGOs (GONGOs), as well as homegrown, grassroots associations. An estimated 200,000 people from all over China, descending on the quake zone, supplying food, shelter, and medical treatment may signal a new confidence in the competency, even obligation of the Chinese people to take advantage of crisis to play a role in assisting the government in relief operations and to engage in the formation of civic associations.[123] Studies of civil society responses to earthquakes and other crises have pointed out the role of crisis in expanding the space for the formation of civil society. In Japan and Turkey, earthquakes forced the government to revise laws governing civic associations.[124] Thus, the Sichuan earthquake can be seen as an opportunity for civic associations to engage in a collective action supported by INGOs and NGOs to develop their competencies.

The definition and discussion of civil society as the associational space located between the government and the nongovernmental parts of society such as individuals, communities, and private and nonprofit organizations should be applied with great caution, since China does not seem to have a civil society.[125]

Commentators often claim that the civic associations existing in China defined as NGOs are supported by the government or controlled by the government to a certain extent. For example, at present more than 400,000 associations are registered with the Ministry of Civil Affairs (MOCA).[126] However, the estimated number of civic associations in China is not supported by reliable assessment, because they include a portion of small "businesses" or unregistered NGOs, which raises the number to 1.5 million associations.[127]

Moreover, restrictive regulations imposed by the government put barriers to proliferation of NGOs in China. Those NGOs that wish to register with Civil Affairs need to meet strict requirements and qualifications with regard to assets, staff, an office, a charter, and so forth. This probably explains why many NGOs in China prefer to be registered by the Industrial and Commercial Bureau as for-profit businesses or to remain unregistered. Given these concerns, civil society in China is always fraught with great skepticism.

Despite these matters, the rapid emergence of grassroots NGOs and networks in China should not be overlooked. Although these associations do not follow the Western definitions and conceptualizations of civil society, it is suggested that in terms of building competencies, these associations manifest a form of self-governance. Even if these civil society groups are formed voluntarily with little or no government support, they cover a wide range of collective activities in the public interest. During the Sichuan earthquake, NGOs, private organizations, and volunteers rushed to the earthquake-stricken region to offer help, or raised funds and materials to be sent to aid those affected by the earthquake.[128] The competencies advanced through the engagement in these NGOs and GONGOs range from "hard" competencies, which refer to more organizational and collaborative capabilities, to "softer" competencies, which refer to creativity and sensitivity. In this regard, hard competencies result in observable behavior, with the invisible, but dominant soft competencies underlying them.[129] It should be stressed that the important role of competencies, both hard and soft, is that they are strongly oriented toward the future. Thus, discussion over these civic associations, whether or not they fit the definition of civil society associations, is of importance whenever strategic community resilience takes center stage in times of great uncertainty. Therefore, in assessing the impact of the Sichuan earthquake on civil society formation in China, we suggest a review of some of the activities taken by several NGOs, INGOs, and GONGOs in creating civic competencies and possibilities for the emergence of civil society in China.

The Sichuan earthquake triggered a large-scale response by the Chinese government and society. One of the primary organizations recognized by the Chinese Ministry of Civil Affairs for conducting a coordinated disaster relief effort, China Charity Federation (CCF), provided urgently needed goods to people in affected areas. The CCF worked closely with the General Headquarters of Quake Relief under the State Council. The China Red Cross was another national-level organization that was quickly responded to Sichuan earthquake. RCSC branches at the provincial and local levels immediately set up local appeals for funding and collected items such as warm clothing to assist those affected by disasters.

Comparative Analysis of Community-Based Disaster Resilience Policies 147

To improve disaster preparedness, RCSC developed a network of six regional disaster preparedness centers providing stocks of relief supplies as well as training staff in disaster management, including logistics, report writing skills, and first aid, tailored to different kinds of disasters.[130] The IFRC's East Asia regional office in Beijing comprises a head of office and technical specialists as well as skilled local staff in disaster management, health and community care, planning, monitoring, evaluation and reporting (PMER), media and communications, and finance. This office supports the East Asia region, which includes China, Mongolia, and the DPRK Red Cross societies in annual and emergency programs. The regional office supports communities in strengthening their capacity to build the resilience of hazard-prone areas in China.

There was also a significant societal immediate response from NGOs, companies, and volunteers in different public domains. In the education domain, several projects were launched to help teachers to continue teaching and improve their competencies in adaptation to disaster events. One of the projects aimed at purchasing sixteen laptops to help teachers in training to proceed with their studies. The project was sponsored by the Aba Teachers College, which is the only college in Aba Prefecture, Sichuan Province, that was severely damaged by the earthquake.[131] Due to the massive damage to school buildings and facilities, the college lacked the tools and facilities to resume teaching. Funds have been provided for computers to fully utilize classrooms and allow the over 4,000 faculty and staff of Aba Teachers College to fully pursue their professional training for enhancing educational reform and minority educational development.

The Beijing Horizon Education Culture Development Center developed intensive training sessions for principals and selected teachers on providing student psychological aid and methods of advancing school resilience in disaster management and postdisaster rebuilding.[132] The principals and teachers were trained in establishing student counseling centers in their schools to provide psychological assistance to students experiencing trauma and psychological stresses. The Chengdu Education Foundation distributed awards to teachers who demonstrated excellence in teaching to improve education for children in affected areas.[133] This foundation was also launched a, five-year program from 2009 to 2013 that provides English communication and basic computer skills to improve the quality and level of education for children in primary and secondary schools in the cities of Dujiangyan, Pengzhou, Chongzhou, and Dayi.

Other projects initiated in the educational domain concentrated on reconstruction of school buildings either completely destroyed or badly damaged and officially declared unusable. These include the rebuilding Guanling Primary School, which is supported by the Beichuan Qiang Autonomous County Bureau of Education and Physical Education in Sichuan Province.[134] Rebuilding the school is of importance as this school has 324 students in eight grades, mainly from the Guanling Village, where Qiang and Tibetan minorities reside. The China Children and Teenager's Fund (CCTF), the first independent nonprofit charity organization in China, established in 1981 by individuals addressing China's children and teenagers' matters, built twenty-three movable plank houses to serve as classroom buildings and facilities to resume teaching.[135] The China Education Development Foundation raised funds for school reconstruction in Ping Wu County, which was heavily damaged and classified as unsafe and unusable.[136] Friends of Nature transported teachers and educational materials to children in refugee camps in Shuangtu, Shuangquan, Maoxian, Mianzhu, Beichuan, and Mianyang County of Sichuan.[137] The Bureau of Education of Jinchuan County, Aba Prefecture, Sichuan Province (Jinchuan Education Bureau) subsidized the reconstruction of three schools in Jinchuan County to benefit 360 children and 20 teachers. The Overseas Chinese Affairs Office (OCAO) Sichuan Provincial People's Government supported the rebuilding of two schools in Mianyang City and Yang Qiao Township.[138] The Pengzhou Municipal Bureau of Education rebuilt the Siwen Primary School in Tongji Township, Pengzhou City of Sichuan Province. The new school was built according to government postearthquake standards, so it can be used as a public shelter for the surrounding community in the event of any future disasters. Large tents were also supplied to serve as temporary schools in Sichuan province until the new school buildings are completed. Organizations also contributed to develop and support public libraries. The CCTF established libraries in sixteen schools and four community settlements for survivors in earthquake-damaged areas of Sichuan. The stated aim of such initiative is to keep children from dropping out of school or from becoming involved in crime. The China Soong Ching Ling Foundation (CSCLF) established seventy-five libraries for rural village and township schools. This project, which Shanghai SCLF titled the "Heal with Love Library Project," benefited sixty schools with a total of 78,239 students in Dujiangyan.

In addition to educational infrastructure, various organizations were engaged in supporting students both financially and professionally.[139] The NPO Development Center Shanghai (NDC) established after-school programs for 6,935 earthquake-affected children that had just moved to boarding schools.[140] The program carries out various activities, including art, sports, reading, writing, and painting to enrich the children's after-school life and enhance their personal competencies and development. The China Foundation for Poverty Alleviation (CFPA)[141] built five sports facilities for earthquake-affected primary schools in poor, rural areas for students to gain a sense of self-esteem and confidence. These students also received psychosocial healing, thereby building confidence to recover from the disaster and rebuild their lives. The Beijing Brooks Education Center developed a culturally competent curriculum for 5,000 students from Chinese minority populations in Maoxian County in Sichuan Province.[142] The curriculum was designed to use local ancestor stories and relevant indigenous customs to help children sustain cultural competency in the wake of the disaster. The Chengdu University of Traditional Chinese Medicine provided scholarships to 147 students affected by the Sichuan earthquake. The scholarship covered academic and living expenses for applicants who demonstrated strong academic excellence. Sichuan University established the Sichuan University Committee of 100 Assistant Scholarships for 78 earthquake-affected students to help students pursue their higher education in the face of the quake's effect on students' academic aspirations.[143]

The support of community-based projects was also shown by various civil organizations. For example, the Asia Foundation, China (TAF-China), together with the Rural Economics Research Institute (RESI) at the Sichuan Academy of Social Sciences, in March 2009 launched participatory planning workshops on resource mapping and community planning involved thirty to sixty participants from each of three targeted communities that have been rated by the State Council as the worst-hit villages in the earthquake.[144] The workshops aimed at helping communities gain access to government relief funds, active involvement in relief programs, and decision making and the finalization of action plans. These workshops hosted local government officials, women's organization representatives, and villagers. The Beijing Pengbo Cultural Training School concentrated its efforts to preserve the Qiang cultural heritage. Since the Sichuan earthquake caused profound losses among the Qiang minority, including large numbers of handicraft masters, professional dancers, and cultural researchers, the Beijing Pengbo Cultural Training School has launched a Help Heal the Soul and Protect Traditional Culture program that trains various Qiang handicraft-making skills to nearly 8,000 students at forty-four local schools in earthquake-affected Beichuan.[145] An inclusion-oriented project was funded by the Asia Foundation, China. In collaboration with Maoxian Women's Federation, Sichuan Shuguang Rural Community Development Support Center, and the Public Health Department of Huaxi Medical College of Sichuan University, TAF-China supported homeless and isolated Qiang women-led households in recovering from the earthquake and taking an active role in community reconstruction and rehabilitation efforts, and thus strengthening their mutual-assistance and networking competency.[146]

The need to restore community mutual support and solidarity in the face of adversity was acknowledged by the Kapok Community Development Research Center.[147] The Center established community centers for relocated communities in the Wudu Village settlement area to enrich Kapok community members through engagement in community events such as dances, concerts, and handicrafts, and community-based volunteer activities. In addition to community-based activities, community members were offered livelihood training workshops in a variety of trades to help them establish their economic competency and start local businesses.

The China Foundation for Poverty Alleviation (CFPA)[148] carried out a microfinance recovery operation plan based on community postdisaster needs. These microloans enable communities to rebuild their local economy and create job opportunities for their members in rebuilding their homes. Other microfinance programs of targeted populations as well as government agencies such as the Postal Savings Bank of China (PSBC) were carried out. The program included business training such as "Start or Improve Your Business" (SIYB) that was jointly organized by the IFRC, ILO, and the consultancy firm PlaNet Finance China.[149]

The Maoxian Association for Development (MPAD) assisted Chashan farmers who lost their homes and source of income, namely selling peppers, vegetables, and fruit, during the earthquake. The MPAD built self-support cooperating groups that include all 135 families in the village engaged in information exchange, cooperation, and community reconstruction planning.[150]

In sum, the NGO response to the Sichuan earthquake is of importance by enabling partnerships between NGOs and local governments, mass organizations and GONGOs. Such partnerships were crucial not only to the effectiveness of the relief efforts but also to mobilize a relatively independent civil society. These civic associations and networks show how China's grassroots associations are capable of forming cooperation and partnerships with the state. The admirable spontaneous and voluntary cooperation in the affected areas could not be facilitated without communities' competencies, such as solidarity and cultural norms of mutual assistance.

Despite the fact that the civic associations' engagement in the relief efforts do not lead to the Western form of civil society, which outlines an oppositional role vis-à-vis the state, their performance symbolizes a central role in building a cooperative sphere within which civil society organizations can pursue their goals even if it is insufficient for a fully functioning civil society.[151] The earthquake had opened up possibilities by providing an associational sphere that enables reducing the long-term government suspicion of NGOs[152] that can lead to decreasing the existing constraints imposed the authoritarian state to a certain extent.

Disaster Resilience Management in Japan: The Great East Japan Earthquake, 2011

The Civil Service of Japan

Although Japan has a long tradition of administrative structure involving the classes of the samurai and the war lords, it was during the Meiji Restoration era (1868–1912) when the Japanese public administration was established in its modern form. Following the Meiji era until the end of World War II, the lower ranks of the samurai class were recruited from graduates of Japan's recognized universities.[153] During the American occupation, the scope of public administration had increased through growing recruitment into the bureaucracy supported by General Douglas MacArthur's indirect rule, which heavily depended on mutual cooperation of civil servants. This was not true for the military service, which experienced limited power and reputation. National government civil servants are divided into two categories, namely special and regular categories. Under the special category are included political or appointments by other factors that do not involve competitive examinations. This category includes cabinet ministers, heads of independent agencies, members of the Self-Defense Forces, officials of Diet (the Japanese parliament), and ambassadors. The core of the civil service is composed of officials of the regular category, who are appointed by competitive examinations. This category is varied, as it includes junior service and upper-professional levels that are also acknowledged as well-defined civil service elite.

The Japanese Society for Public Administration (JSPA) was established in 1950, stipulating public administration as a discipline. The JSPA was supported by officials and law students who became involved in academic public administration research domestically and through cooperation with Asian and other specialists and institutions. Japan's high-speed economic growth in the 1960s was due to the MITI's role of administrative guidance, which gained international recognition by the publication of Chalmers Johnson's book *MITI and the Japanese Miracle*.[154] The book described the central bureaucracy of the Ministry of International Trade and Industry as it worked out a firm alliance between bureaucrats and politicians while at the same time recognizing that bureaucracy needs to work in politically neutral ways.

During the late 1980s, bureaucracy continued to play an important role in Japanese policy-making, while declining birth rates and the pressure to increase Japan's economic performance appeared to have a growing impact on bureaucrats' power.

In the face of such pressures, it is argued that the central government bureaucracy will continue to play a significant role in the Japanese state and society, and the public service employees still enjoy many of the benefits introduced in the boom years.

At present, the Japanese civil service has over one million employees, with half a million employees in postal service, or Japan Post (since 2003). Although growing privatization processes of government agencies and institutions such as the postal service is currently on the government agenda, the government is still recognized as an employer of great importance.[155] After the breakdown of the Japanese asset price bubble in the early 1990s, wages and privileges in the private sector were cut, but public service workers still enjoy many of the benefits introduced in the boom years. In the face of the 2009 general election, the Democratic Party of Japan (DPJ) came to power after the prolonged regime of the Liberal Democratic Party of Japan (LDP). The DPJ has offered to set up a reform referred to as "leadership by politics" that aims at challenging the civil service practice era of the LDP and offer new reforms of the civil service.

Advocacy

At the time of this writing, Japan is still suffering the economic and environmental impact of the Fukushima Daiichi accident. Progress appears to have been made in the management of the damaged reactors and fuel, and the only deaths reported as a result of plant operations were not caused by nuclear radiation; immediate physical effects from the earthquake and tsunami resulted in victims.[156] Comprehensive and in-depth assessment of the nuclear accident and its effects on the plant's workers, the public, and the environment will likely take years. Furthermore, validated information from the plant is difficult to gather, since response and recovery operations are still ongoing in the tsunami disaster area. However, the immense effects and damage caused by the nuclear reactors and the interplay of those failures with emergency management efforts have already had significant impacts on the public consciousness and community advocacy in particular.

Despite the fact that this was the most photographed nuclear accident in history worldwide, residents of the affected areas were not kept well-apprised of developments and often received vague, confusing, information that is not up-to-date. The media, however, is not the only party that acted dysfunctionally and ineffectively in communicating to the public. The government tried to fill the information vacuum in the media, frequently through populist statements and speculation. For example, during the second week after the disaster, the death toll reached 10,000. Prime Minister Naoto Kan addressed the situation at the Fukushima plant as "grave and serious," and added, "We are not in a position where we can be optimistic. We must treat every development with the utmost care."[157] Kan tried to use a submissive approach by apologizing to the farmers and business owners around the plant for damage caused.

These statements seem rather poor and indefinite as a response to the criticism that the government has failed to communicate in a clear and effective manner with the public and foreign governments about the situation at the Fukushima plant.[158] Moreover, the government showed lack of leadership and largely failed to manage the crisis effectively. It was criticized for not being fully transparent; not heeding warnings; not establishing an independent regulatory agency; not providing substantive and clear disclosure on issues related to the nuclear disaster, such as why it took five hours after the first reactor explosion for the government to report that no radioactive material had been leaked, the amount of spent fuel reactors contain, or how these threats affect the public; and lack of government and TEPCO assessment of how to prevent further damage and devastation.[159]

For affected communities to take actions to advocate for themselves, appropriate and reliable knowledge is crucial. At that moment, expert nuclear professionals (and their employers) in government, academia, and industry avoided the press and did not provide the public with facts and useful guidance. Even if the government and Tokyo Electric Power Company (TEPCO) have given numerous interviews and press conferences, the information was poorly delivered and made overuse of engineering and scientific terms and assessment of the events at Fukushima Daiichi that go beyond common people's understanding. These proved to become barriers to developing advocacy, together with lack of support from institutions, government, or both. The lack of support for advocacy was also shown by the reluctance of the government to support the affected communities and the public in general in articulating their views, motivate the expression of concerns, and opinions to allow them to make their own assessment of risks and choices.[160]

Several advocacy efforts were made to press the needs for adequate information in the face of the disaster and nuclear adversity in Japan. On March 26, twenty-five Diet members signed a letter calling on the government to send young children and pregnant women out of the thirty-kilometer danger zone around the damaged Fukushima No. 1 nuclear power plant.[161] The statement required extending the existing twenty-kilometer mandatory evacuation zone radically to avoid further exposure and possible long-term radiation harm, arguing that

> A grave situation continues at the Fukushima No. 1 nuclear reactor, while at the same time there is no order for an expansion of the

evacuation area. Reactor No. 2's containment has been damaged, there have been fires at the Reactor 3 and 4 spent nuclear fuel pools.[162]

The statement also criticized the misleading information delivered by the mass media and by the government that there is no immediate harm to human health, as such assessment ignores the possible long-term damage. The statement was signed by members of the Democratic Party of Japan, including Kazumi Ohta, a Lower House representative from Fukushima Prefecture, members of the Social Democratic Party, and one from the Liberal Democratic Party. Antinuclear activist groups marched through Tokyo on March 27, after they advertised the rally on social networking sites as workers battled to stem the damage from a leaking atomic reactor. They held the march to encourage more citizens to sign the statement.

Other sporadic efforts were held by the local officials. The breakdown in confidence also grew among local mayors. The local mayor of a town close to the Fukushima Daiichi nuclear complex did not hesitate to criticize the central government for not keeping his office updated on the situation. According to Katsunobu Sakurai, mayor of Minamisoma, "We've been asking the prefecture and the government to give us information quickly but we've been having to force information out of them."[163]

Some local media have also joined to adopt a critical tone of the government and Tokyo Electric Power Company for failing to keep the public well informed on the developments. A Mainichi newspaper editorial expressed its disappointment of the disclosure on the nuclear accident.

> Information is the essence of crisis management . . . based on real-time information, it is vital for the government to join as one with experts in nuclear power and radiation, crisis managers and experts in public relations and risk communication to work to make information available.

The *Yomiuri* newspaper harshly criticized the prime minister's administration for raising public anxiety due to a total lack of government plan: "Televisions reported an explosion. But nothing was said by the premier's office for about an hour."

Social media such as Twitter helped citizens gather information on the government and TEPCO actions. Moreover, government officials such as nuclear safety agency officials and Fukushima plant employees acknowledged that they suspected a breach in the reactor core of one unit at the quake-damaged plant. However, these officials were warned by a local official to stop discussing what they had seen.

To conclude, the implications of recent catastrophic disaster, the Fukushima Daiichi nuclear power plant accident, reach well beyond the immediate, direct environmental and human health risks. The government's lack of transparency in managing this crisis imposed barriers on community advocacy, threatening to undermine public confidence as a whole.[164] It has been argued that communities at risk should be well informed in order to assist relief and recovery efforts by drawing on their experiences in overcoming disadvantage and how their experiences are embedded in the multiple and wider aspects of their lives. These insights may promote positive adjustment in adverse circumstances. Thus, in order to determine if and when a community "is doing fine" in the face of adversity, the affected community has to be asked about how it is experiencing the situation. The voices of those afflicted need to be heard if we are to understand their experiences of adversity, and to identify solutions that fit the constructions of reality. The need to keep citizens well apprised of risks can be seen in the dramatic events of twenty-five years ago at Chernobyl. In the aftermath of the Chernobyl disaster in 1986, the Kremlin failed to inform the public of the disaster. The information about the disaster was delivered by Swedish scientists who detected high levels of radiation.[165] The implication of such failure reached beyond the specific accident and actually led to the collapse of the Soviet Union more so than the launch of perestroika (restructuring). It is suggested then that in times of disaster, government needs to become an advocator by creating an environment that is open and supportive to decision making.

Inclusion

The earthquake in Japan and the Fukushima Daiichi nuclear power plant accident have led governmental, international, and NGOs to use targeted interventions to groups most at-risk such as children, women, and older people. Save the Children estimated that as many as 100,000 children may have been displaced during the disaster. Some were separated from their families because the earthquake and tsunami struck in midafternoon when many children were at school or nursery. The Japan Asian Association and Asian Friendship Society (JAFS), an NGO involved with international cooperational work and networking, offered story-telling programs for children by using *kamishibai*, a picture card show, and footbaths to help reduce the psychological stress and trauma of children in affected communities located in Kaikan in Utatsu District, Minamisanriku Town, and those who live around the Minamisanriku Town in Miyagi Prefecture.[166] On May 5, JAFS organized special activities for children to celebrate Children's Day in Japan. ICAN, an NGO located in midland Japan, supported children and their parents living in evacuation centers in Ishinomaki City, Miyagi Prefecture.[167] Among its relief efforts, the organization provides food for the families, school supplies for children, and organizes movie events, especially for children living in evacuation centers. The NGO Good Neighbors JAPAN aims at providing support for children located in Kamaishi City, Otsuchi Town, Yamada Town, and Iwate Prefecture.[168] The organization sent volunteers to visit kindergartens and nursery schools to deliver commodities, to clean mud, and to construct prefabricated kindergartens, as well as provide mental health care by psychotherapists. It has also arranged two events called "Mothers and Children's Flea Market" (in May) and "Let's Do Otsuchi Reconstruction Festival" (in June) to allow children and mothers to regain their daily activities and functions and cheer themselves up. Another NGO, called the Terra People Act Kanagawa (TPAK), offered mental health services for 140 children, their families, and their teachers in two nursery schools in Kamaishi City in Iwate Prefecture.[169] The volunteers initiated cultural activities with the slogan "Reconstruction with brilliant smiles of the children," as well as provision of goods and education for children of the two collapsed nursery schools in Kamaishi City. The Terra People Association NGO that supported affected children located in Ishinomaki City and Kesennuma City in the Miyagi Prefecture, came up with a project named "The Festival for Life" to connect the affected children with children around the world to bring them happiness and hope [170] The project aimed to help the children affected by the earthquake remake their dreams and hope for the future with children all over the world. The NGO Child Fund Japan help deprived, excluded, and vulnerable children by organizing mental health–care programs for children and workshops for training elementary school teachers by providing immediate mental health care in Tokyo, Fukushima Prefecture, and Iwate Prefecture.[171]

The World Vision Japan, a Christian relief, development, and advocacy organization, worked with children from Minami Sanriku Town, Tome City, Kesennuma City (Miyagi Prefecture), Yamada Town, Tono City, Otsuchi Town, Miyako City, Tanohata Village, Noda Village, Kuji City, and Ichinoseki City (Iwate Prefecture).[172] The organization had managed Child Friendly Space (CFS) to supply mental health care for children, and school aids, buses, and lunch to support school life for children. The Organization for Industrial Spiritual and Cultural Advancement–International (OISCA), a Japanese NGO, developed a project to focus on psychological support for children through wooden building blocks that it established as educational tools to develop environmental and emotional education in the Northern part of Ibaraki Prefecture, Iwaki City in Fukushima Prefecture, and affected areas in Miyagi Prefecture.[173]

Other inclusionist projects targeting both children and the elderly were implemented by the Action against Child Exploitation (ACE). The organization, an NGO, provided education, livelihood, and local activities to reduce mental stress and trauma in children, junior high school and high school students, and elderly people who live in the temporary housing in Yamamoto town and in Sendai city.[174] The organization planned a Kakigori Festival two months after the disaster to encourage young and elderly participation in daily activities on the community level. The volunteers also built a playground for children. The ADRA Japan, an NGO, targeted elderly people in nursing homes in Sendai City and Higashi-Matsushima City in Miyagi Prefecture, Yamamoto Town Hall, residents in Yamamoto Town, and those who live in temporary housing in Higashi-Matsushima City, Yamamoto Town, and Watati County.[175] Its main initiative is to offer daily meals and other necessities (e.g., dishes, bedclothes, cooking materials) for elderly people's houses.

An organization with a wider focus of targeted communities at risk is the Caring for Young Refugees (CYR). This NGO offered to support young children, their mothers, and staff members of nurseries in Miyagi and Fukushima prefectures. It provided child-care service activities in plastic tents or in evacuation centers in thirty project areas. It also created child-care kits that included picture books, paper and clay dolls, and cakes that are selected by specialists in child care and psychology for child survivors in affected areas.[176] Oxfam Japan offered various services for small children and their mothers, single mothers, people who have immigrated to Japan, and people who faced domestic violence and sexual violence during the disaster in areas where victims are evacuated, such as Morioka City (Iwate Prefecture), Sendai City, Tagajou City, Shiogama City (Miyagi Prefecture), Koriyama City, Aizu Wakamatsu City, Fukushima City (Fukushima Prefecture), and Hitachi City (Ibaraki Prefecture).[177] To assist these marginalized groups to participate in community activities, the organization provided information, counseling, and psychological support for female survivors in order for them to access the government social security program.

The Japanese Organization for International Cooperation in Family Planning (JOICFP) has joined other organizations such as Japanese Midwives' Associations, Japan Family Planning Association, and Oxfam Japan to meet the needs of women, pregnant women, and new-born children who were affected by the disaster.[178] Among the provisions offered to women are emergency relief supplies for women, pregnant women, and newborn children, aid for local midwives' activities, access to maternity hospitals inside or outside of the prefecture, family planning services, and financial aid for pregnant women (provision of fifty thousand yen). The Japan Association for Refugees (JAR) offered assistance to marginalized groups such as women and foreigners in both Iwate and Miyagi Prefecture.[179] Its relief efforts include communicating with refugees, providing a "women's kit" considered beneficial for women, preparing guidance with midwives, and providing legal aid.

It should be noted that the main activities in response to the large-scale disaster were provided by governmental agencies, such as the Overseas Disaster Assistance Division, the Economic Cooperation Bureau of MOFA, the relevant divisions of ministries and agencies concerned, and Disaster Assistance Division of JDR Secretariat of JICA. The JDR teams dispatched to disaster-affected areas included teams of rescue personnel from the National Police Agency (NPA), the Fire and Disaster Management Agency (FDMA), and the Japan Coast Guard (JCG). Medical teams of medical doctors, nurses, and medical coordinators registered with the JDR Secretariat of JICA. Expert teams of professionals and experts in particular fields, such as emergency disaster management and disaster restoration, were sent from the related government ministries and agencies of Japan to the affected areas.

However, NGOs' targeted interventions during and after the 2011 disaster in Japan were more able to single out communities with an increased probability of showing maladaptive adjustment, yet it might miss a considerable number of groups where no problems were yet apparent, such as teachers, including kindergarten teachers, and other community members who are less identifiable problem groups with fewer shared characteristics. Volunteers from the Israeli NGO IsraAID (interview July 2011) established child-friendly spaces run by local volunteers and equipped with interactive toys, arts and crafts materials, and games. One Israeli volunteer reported that teachers and assistants in day-care centers shared with her their feelings, such as being helpless and blamed by the community members for not protecting the children during the earthquake.

Because human development is changing and dynamic, disadvantaged community members have the capacity to move in and out of extreme situations. Therefore, the narrow focus of targeted policies creates concern about the availability of support for all when needed and for all who can benefit. Access to services should be facilitated for all, aiming to raise the floor for every person, at every age, and in every place. In this way, community inclusion can lead to active participation in community engagements and decision-making.

Competency

Building a resilience framework entails emphasis not on deficit but on areas of strength, building up resources from inside the community care network, and strengthening the social fabric. Identifying and building on the strengths of community members in adverse circumstances can promote their own feelings of competence and capability and can stimulate enduring positive changes.

Several efforts were made by NGOs to harness momentum for positive adaptation and recovery in affected communities. In the domain of livelihood and housing, Bridge Asia Japan (BAJ), a nonprofit organization helping socially vulnerable and disadvantaged people (refugees, disabled people, women) in the areas of Ofunato and Rikuzentakada City, Iwate Prefecture, provided various nutritious dishes by local chefs, whose restaurants were damaged during the disasters, on a volunteer basis.[180] The Institute of Cultural Affairs (ICA), an NGO, collaborated

Comparative Analysis of Community-Based Disaster Resilience Policies   161

with municipal offices in Miyagi and Fukushima prefectures in order to help the local fishery cooperative association rebuild their offices in their harbor so they can restart fishing operation in a few months.[181] The International Volunteer Center of Yamagata (IVY) established the Tohoku Wide-Area Disaster NGO Center to provide assistance to victims living in shelters in remote locations and small-scale shelters in locating employment opportunities under the program as "Cash for Work."[182] Other organizations offered more basic and daily needs to encourage victims to improve living conditions by themselves. Habitat for Humanity Japan collected donations for people affected by the Great East Japan Earthquake to meet communities' needs assessment with regard to goods, tool kits, and commodities necessary for rebuilding of houses and reconstruction of communities.[183] The Peace Winds Japan (PWJ) is an NGO affiliated with the Chamber of Commerce and Industry that plans sustainable and culturally appropriate recovery solutions for both immediate relief from disasters and longer-term restoration.[184]

In the medical domain, NGOs endeavor to change the way medical agencies interact with communities and the way they both plan to deliver health-care services. The Association of Medical Doctors of Asia (AMDA) has provided medical relief since March 12, 2011 in Minamisanriku Town, Kamaishi, Ohtsuchi Town (Iwate Prefecture).[185] Since April 20, 2011, AMDA has moved to supply reconstruction support and local medical facilities. Efforts were made by the organization to support educational assistance by establishing scholarships for the survivors of the disaster, and a health-care center. The Service for Health in Asian and African regions (SHARE) is an NGO that provided medical services for the victims who moved into temporary houses.[186] The organization collaborated with local figures and medical entities to implement inclusive health and long-term medical support. SHARE volunteers visited constructed temporary houses and participated in a health consultation unit that was formed under the cooperation of a medical support team from outside the prefecture.

Japan Overseas Christian Medical Cooperative Service (JOCS) and the Médecins du Monde Japan offered immediate medical care assistance at evacuation centers with psychiatrists, nurses, specialists in exercise therapy, clinical psychotherapists, psychiatric social workers, acupuncturists, medical coordinators, general coordinators, and doctor and nurse managers in Arahama District, Sendai City, Otsuchi Town,

Iwate Prefecture, by supporting the Tohoku District Center for victims of the United Church of Christ in Japan.[187]

Through education provisions, organizations and volunteers aimed to build a positive community-school relationship around development of community competency. The *Kokkyo naki Kodomotachi*, Children without Borders (KnK), coordinated with boards of education in five municipalities in the coastal part of Iwate Prefecture (Yamada Town, Otsuji Town, Kamaishi City, Ofunato City, and Rikuzentakada City) regarding educational facilities, school transportation, school uniforms, and equipment for staff rooms in order to reopen and to proceed with education at elementary and junior high schools.[188] The organization worked closely with the cooperation of local communities to support appropriate mental health care for children by arranging sports and cultural activities. The JEN organization established programs to encourage community members in Ishinomaki City, Iwate Prefecture, who lived in evacuation centers and houses in Ishinomaki City, to meet with each other by providing spaces for community gatherings such as body care services (massage, makeup, hairdressing), entertainment activities in collaboration with fashion magazines and artists, and business recovery workshops.[189] The Association for Aid and Relief, Japan (AAR) provided necessary daily commodities for all who live in temporary shelters and psychological care for elementary and junior high school students who are affected by disasters. In addition, the organization offered long-term support for elders and people with disabilities.[190]

NGOs also placed emphasis on enhancing the cultural competency of affected communities. For example, CARE International Japan arranged community events in affected areas, such as cherry blossom viewing parties and events for children to motivate local communities to rebuild and to share local information. It also supported the publication of a community newspaper in which there is necessary information for evacuees.[191] The Japan Team of Young Human Power (JHP) established a disaster volunteer center in collaboration with local social welfare associations in Minami Sanriku town, Miyagi Prefecture. The organization helped to facilitate the registration and assignment of volunteers as well as needs assessment of evacuation centers and affected communities.[192] The SHAPLA NEER Citizens' Committee in Japan for Overseas Support collaborated with other NGOs and the Fukushima

Prefecture and Social Welfare Council in Iwaki City that worked in Iwaki City Fukushima Prefecture to support establishing volunteer centers.[193] The Japan Volunteer Center (JVC) managed support for victims in Miharu Town in Tamura County, Fukushima Prefecture, especially farmers' communities through the Kessensuma City Volunteer Center set up under the Council of Social Welfare of Kesennuma City in order for local people to play a leading role in reconstruction and recovery activities.[194] The organization also distributed radios for the people in cities who have been seriously damaged by the effects of the nuclear accident to attain necessary information via a newly set-up special disaster broadcasting station.

The Shanti Volunteer Association (SVA) offered support for local communities ("town development" support activity) by arranging meetings between residents for needs assessment and planning of community festivals/events such as tea parties to promote town development and reconstruction efforts. In addition, the organization established libraries in local community centers.[195]

These interventions attempted to improve isolated skills or competences with consideration of the wider context in which they occur—that is, the tsunami, earthquake, and nuclear disaster that occurred in Japan in March 2011. The NGOs aimed at meeting the functional utility of the competencies targeted within the disaster that reinforces and maintains them. However, intervention or prevention programs should be integrated into the cultural context of affected communities in order to promote a sense of affiliation, value structuring, achievement, and experience of mastery to remove obstacles to the acquisition of capabilities in the face of future adversity.

7
Administration and Community Collaboration in Disaster Management

Professional Helpers in the Service of Disaster Resilience

The aim of this chapter is to take a more practical step in striving toward the ideal of professional helper in practice. The focus of the chapter is on public administrators and what they need personally and professionally to be the best helpers possible in the face of adversity. Thus, this chapter introduces public administrators to the attributes of an effective helper. There is no one set of characteristics that identifies professional helpers, but we encourage pubic administrators to reflect on the characteristics they possess that can either help or hinder them in their resilient efforts and work with affected communities. This chapter thus underlies the need for professional helpers to consider that their personal cultural network is the starting point for how they engage disaster events and for the methods they use in their professional work.

The theory of communitarian-based resilience management and the analysis that followed seem to establish a normative model and distinguish it from others. Communitarian-based resilience management is difficult to establish without linking community and administration. A comprehensive community perspective on resilience management involves three levels of intervention: (1) advocacy, (2) inclusion, and (3) competency. These attributes emphasize deep social and community transformation rather than merely helping people adapt to risks and crisis circumstances.

To embrace these roles, professional helpers need to acquire distinguished professional ethics based on understanding themselves as cultural beings. Culture is referred to here as insertion in a community as a means of belonging. Culture is a way of life, a group of beliefs, traditions, and techniques. Culture varies according to communities, societies, or ethnic groups.[1] Public administrators become cultural beings by learning to participate in the cultural activities and practices going on around them. Tomasello[2] defines cultural abilities as a uniquely human social-cognitive skill for understanding others as intentional agents who, like the self, attend to things and pursue goals in the environment.

Professional helpers need to consider that their personal cultural network is the starting point for how they engage disaster events and for the methods they use in their professional work.

Professional Helpers as Advocates

Professional helpers must meet the physical survival needs of the affected communities from the outset. In times of emergency, victims often do not have the luxury of seeking alternative options for obtaining needed goods and services.[3] This is by no means a call for establishment of client/customer and business relations that respond to short-term interests of customers served by government programs. Public servants need to go beyond short-term interests and to assure that disaster response and recovery efforts are consistent with norms of justice, fairness, and collective shared notions of public interest. Treating communities at risk as clients degrades their effectiveness, commitment, and responsibility for what happens in their communities. Therefore, public administrators need to respond not just to customers' needs, but focus on building relationships of trust and collaboration with and among community members. In times of disaster, the primary task of professional helpers is to "serve" rather than to steer by empowering citizens to articulate and meet their shared interests to enhance immediate coping capacity. The role of professional helper is to assist vulnerable people in identifying a network of resources available to them, such as family, friends, and community. Professional help needs to connect the affected individuals to community resources such as schools, churches, and other natural support systems existing in the community. These community agencies provide an initial safety network based on the sense of social connectedness needed in times of crisis. Arranging such connections creates the opportunity for vulnerable people to learn what the community has to offer them in the future, to manage their own lives in the face of adversity. Engaging in community resilience activities may serve not only to intervene in disaster management but also to prevent people from entering into a poor state of functioning at the initial stage. Communities with accessible agencies and services designed from the bottom up offer opportunities to become involved there rather than drowning in acts of despair.

Thus, for the initial encounter with communities at risk, professional helpers need to help in a manner that reflects maintaining personal networks with community members who may be in a position to assist vulnerable people in making an assessment of the immediate situation, especially of their coping resources. The following are tasks that public servants might be involved in when advocating for vulnerable communities:

- achieve credibility and recognition within communities at risk;
- gain meaningful knowledge of communities' services and agencies that could provide immediate service to vulnerable individuals;
- establish a personal network of community leaders and members who may be in a position to give immediate assistance;
- keep vulnerable people involved in resilience efforts connected to each other; and
- assist the community to articulate its needs in a way that enables the government to act on them.

Professional Helpers and Inclusion

In times of disaster as well as in daily practice, ethnically diverse communities, individuals, and communities in rural areas, older people, women, foreigners, and children would not consider seeking government resilience programs.[4] This could result from feelings of insecurity, distrust, and inconvenience with civil servants who are not members of their help network. Often they may turn to gain assistance from family members, friends, and ethnic community members. Thus, professional helpers need to encourage community leaders and agencies to acknowledge their members with the full resources existing within their own communities, including community centers, religious centers, extended families, neighborhood social networks, and ethnic or women's advocacy groups. Public servants are then required to assume an open stance to various local practices of relief that exist within excluded communities' culture. Such knowledge of local practices could be addressed by engaging with leaders or spiritual healers of the communities at risk. In addition, policy implementation should not guide vulnerable communities in the "proper" direction. Public officials need to collaborate with NGOs and other community agencies to seek solutions or appropriate policies to the specific problems these communities face. Viewed in this way, professional helpers may play more than a technocratic function of delivering goods and executing government regulations, but rather serve as mediating actors assuring that resilience efforts are consistent with the special vulnerabilities possessed by excluded communities, both in substance and in process. Strategies to enhance the inclusion of local communities in administrative relief efforts and decision making include the following:

- becoming aware of differences in cultural definitions of risk, health, and security;
- becoming aware of biases and of sociopolitical issues that impose difficulties and barriers to community members from diverse backgrounds;
- implement resilience services without discrimination or preferences based on age, gender, ethnicity, disability, race, or socioeconomic status;
- actively reach out to minorities and groups with special needs and initiate resilience efforts to create more long-term solutions rather than merely treating problems;
- negotiate and consult with community leaders and NGOs about local relief activities;
- work with members of minority communities to develop and build on community resources to promote their self-resilience with shared decision-making authority;
- make sure that community needs and activities are reflected in agency budgets; and
- develop training programs to prepare community members to assume proactive roles and leadership skills to influence policy makers.

Professional Helpers and Competency

The role of professional helpers is to help citizens articulate and meet their shared interests and to use their own competencies to create collaborative partnerships with community members, which means that they assume a fair share of the responsibility for resilience management policies. Professional helpers need to acquire listening and empowering skills to assist communities in becoming aware of their capacities and strengths rather than concentrating on barriers, problems, and risks. Communities need to feel officials' respect and empathy toward their ability to exercise control over their lives, rather than stereotyping them. In addition to demonstrating respect, professional helpers need to put effected communities in touch with external resources within the community and beyond (such as government agencies) to cope effectively with the disaster and with day-to-day survival issues such as finding a job, arranging for child care, and so on.

Inviting communities to consider the strengths they possess, but may not be using, requires focusing on concrete problems while dismissing sweeping judgments. The emphasis is on asking communities to come up with specific plans for what they did before and during the disaster to attain better adaptation and resilience in the face of adversity. Even in times of disaster, the critical element is to enhance local competencies based on internal and external resources and limitations to ensure these strategies are specific, realistic, and best suited to their capabilities and value system.

- Take steps to avoid fostering dependency for communities so they can assume responsibility for action.
- Emphasize positive and constructive thinking patterns to help communities discover their own resources and competencies.
- Explore a range of opportunities for resilience and coping with risks so that affected communities can select which option is best suited to their capacities and shared values.
- Develop training workshops for effected communities to learn about resources and capacities that are available to them in order to develop plans that they can carry out independently.
- Create specific and concrete resilience plans necessary for community members to commit themselves and mobilize their resources to a definite plan they can realistically accomplish.

Conclusion and Guidance for Public Officials and Community Collaboration in Disaster Management

The starting point for writing this book was the urgent need for a new equitable and effective emergency management program aiming at a resilience approach based on social justice when applied at the level of interaction between administration and affected communities.

Community-based resilience management is presented here as a dynamic process of positive adaptation in the face of significant adversity or trauma underlying mechanisms and processes that enable community members to develop and maintain their shared interests and community values, including advocacy, inclusion, and competency, despite the experience of adversity.

In this book I have reviewed community-based strategies in order to understand how disaster relief efforts can best be provided to enhance resilient communities. This study provides an insightful understanding regarding the relationship between public administration and community resilience. The central role of public administration lies in developing empathetic relationships and facilitating collaborative action and better resilience outcome. However, it is suggested that the ordinary mode of disaster administration cannot address such goals. This might be because the resilience process requires well-informed, confident, and active participant communities rather than passive receivers. Resilience can play a crucial public role in helping institutions to empower citizens to exercise their shared interests and trust in the efficacy of collaborations with public administration to mitigate the effects of future natural disasters on society as a whole. Public administrations are then responsible for assuring disaster response and recovery, efforts are fully consistent with communitarian norms and justice. Thus, administration and community interaction in enhancing disaster resilience needs to be examined in terms of social justice, which adds a new dimension to the evaluation of disaster management policies.

By using social justice theories to address concerns of disaster-affected communities, it is suggested that Walzer's communitarian social justice theory helps one to envisage the political process as generating a range of provisions that express a shared concept of social citizenship through the gradual formation of legitimate expectations concerning the access to needed goods in the "society of equals" as a whole. Thus, incorporating principles of disaster resilience in emergency management policy should not just be phrased in the introductory sections of the public administration Code of Conduct or ethical codes; they need to guide disaster resilience policy in an effective and equitable way, as these principles carry with them social justice standards of action and provision.

In this book I explore the applicability of a disaster-resilient community model that meets the criteria of social justice in various regional settings. Although comparing these four events seems odd, since the cases of the Gulf Coast Hurricanes (United States), the West Sumatra Earthquakes (Indonesia), the Wenchuan Earthquake (China), and the Great East Japan Earthquake strongly differ in their socioeconomic development, political regime, and cultural attributes, the Indonesian and Chinese cases have proven that their administrations were able to respond immediately to natural disaster. The most problematic dimension of community-based disaster resilience is community advocacy. The comparative analysis has shown that public officials did not act as advocates for New Orleans's poor communities in the case of urban renewals and educational reforms in post-Katrina New Orleans. Administrators paid little attention to the meaning of *house* or *home* as perceived by residents in this particular situation and, therefore, did not contribute to communities' advocacy actions in providing appropriate responses to survival needs. The right to housing was not fully appreciated by public officials and the government in general, as housing is more than a safe place, but rather a place marked by familiar locals, social networks, and cultural norms and practices. The inherent class and social vulnerability before Katrina contributed to the weakened capacity of New Orleans residents to articulate and meet their shared interests.[5] This was also the case in Indonesia, as the lack of efficient and valuable advocacy efforts on the part of both public officials and communities was intensified by the fact that communities became involved in choosing settlement and shelter options, resulting in prioritizing the external appearance of masonry houses to represent modernity and affluence associated with middle class rather than greater safety building standards. The lack of professional and skilled administration was intensified in the Indonesian case, where communities were left alone to decide on security and housing matters with no collaboration and engagement with government agencies. This case proves that the power of communities and the capacity of public administration can increase together, in a positive sum interaction, or may also decline together, in a negative-sum way, as when the state's administrative capacities stagnate along with communities' capacities for independent self-determined activities.

In China advocacy initiatives were mostly suppressed by the government. The Chinese government has expressed a zero-tolerance attitude toward the public criticism that was raised over the school collapse during the Sichuan earthquake. The school scandal in China signified a growing public distrust of government and administrative responses. Despite a successful campaign to silence parents and social activists, parents' groups were able to translate their pain and anger (as well as expectations) into sporadic collective actions that speak to the economic and political concerns of the government. In Japan, the nuclear accidents at Fukushima Daiichi were poorly handled by both the government and Tokyo Electric (TEPCO) in supplying appropriate and reliable knowledge needed for affected communities to take advocacy actions to make their own assessment of risks and choices of response. It should be added that sporadic efforts did help local officials to stress the need for adequate information in the face of the disaster and nuclear adversity in Japan.

In the realm of community inclusion, NGOs and local New Orleans organizations aimed at utilizing neighborhood collaborations and participation in local decision-making. However, the case of school reform shows that exclusion of poor communities has not yet been resolved but rather intensified. NGOs targeting inclusionist interventions during and after the disaster in Indonesia and Japan were more able to single out vulnerable segments of society with an increased probability of showing maladaptive adjustment, such as children and women, yet it missed a considerable number of groups whose problems were not yet apparent. More successful inclusionist projects were evident in China, such as those of the Qiang women and peasants, who actively engaged in Sichuan community activities and gained a sense of belonging. However, such projects relied heavily on government initiatives, especially in autocratic regimes, and should not be seen as a long-term commitment of government agencies to change existing regulations and standards because of unclear future public interpretation of new regulatory rules. This may result in the unfair burden of disaster risks and vulnerabilities borne by a minority community. The dimension of inclusion thus outlines the contradictory role and strategies regarding community inclusion in the face of adversity. On one hand, government officials are expected to represent the best interests of effected communities. On the other hand, when state government officials and agencies ignore local and at risk community complaints and concerns, these communities can turn to protests to have their voices heard in court and by the government, fueled and supported by foreign, media, and human rights advocacy coalitions.

In contrast to advocacy and inclusion, community competency was shared by these four countries. The effort to recover from Hurricane Katrina seems to have spurred the growth of communities' competencies in New Orleans, especially in ethnic and solidaristic communities such as the Jewish community, the Vietnamese community, and the Latino community. Sadly, in too many cases, officials got in the way of nonprofit, community-based organizations and activist groups, waving red tape and rule books. A similar situation was also evident in Indonesia, where community competency was intensified during the recovery phase. The new reforms issued by the Indonesian government, weakened the traditional functions of the *Gampong*, meaning the citizens could not use their power to prevent the development projects that entered remote areas in North Aceh and in the Gayo highlands, leaving the *Keuchik* and other *Gampong* members with no income from their *seunebok*, or productive land, which was different from what had been the case in the past. Thus, the *Gampong* communities became poorer and their political participation had been weakened. In both China and Japan, recovery efforts greatly relied on community competencies. The earthquake had opened up possibilities by providing an associational sphere that enabled reducing the government's long-term suspicion of NGOs, as NGO response to the earthquake is of importance by enabling partnerships between NGOs and local governments, mass organizations, and GONGOs. Such partnerships were crucial not only to the effectiveness of the relief efforts, but also to mobilize a relatively independent civil society. The admirable spontaneous and voluntary cooperation in the affected areas could not be facilitated without communities' competencies, such as solidarity and cultural norms of mutual assistance.

In terms of the normative model we examine here, public administration ethics and practice seem most consistent with the basic foundations of democracy and, therefore, provide other valuable functions of public administration in the face of adversity, fully integrated with community resilience and public interest. As seen, the role of administration and community participation is crucial for greater enhancement of disaster resilience efforts. In times of emergency, in the course of assisting and rescuing victims, and transporting them to nearby medical and shelter facilities, the relationship between public officials and affected communities should not overlooked. It is important that public officials as professional helpers know how families and communities in different societies adapt to cope with the impact of risks and disaster situations. This is a result of a socialization process that utilizes and makes effective the operation of these relationships before, during, and after disaster. This book has highlighted that family and community competencies in times of disasters have an important role in advancing disaster resilience. Therefore, the various ways in which public officials use communities as transfer agents, acknowledging their differences in vulnerabilities and behavior or coping strategies in the face of disaster, empowers the communities. Sharing information, cooperation, collaboration, and involvement in communities as well as with other stakeholders such as NGOs, agencies, and departments paves the way to the development of disaster-resilient communities and civil society as a whole.

Drawing on the findings from the comparative analysis, the design of community-based disaster resilience policy needs to meet the following key guidelines:

Guidelines for Public Administration

Advocacy

- Create official understanding of community goals and vision by visiting community institutions (child-care centers, hospitals, schools) and holding regular meetings with community representatives.
- Official reports on disaster risks and implications should be carried out to include both professional risk assessment and community risk assessments that are articulated by community members.
- Make sure that updated disaster risks, perceptions, solutions, and findings provided by the affected community will be fed into community resilience planning.

- Discuss with community representatives the appropriate design of awareness-raising programs at the community level.
- Use local and community knowledge to discover critical facilities of the community (e.g., shelters, evacuation, communication, first aid centers, supplies).

Competency

- Advise community leadership or representatives of community and individuals' rights and the legal and regulatory obligations of the government and its administrative agencies to provide protection and recovery in disastrous events.
- Provide professional counseling to advise the community on how to manage its resources and material assets.
- Promote community and external agencies and organizations partnerships by encouraging them to participate in disaster preparedness and assessments decision-making.
- Incorporate maximum utilization of community resources and capacities in preparedness and recovery official policies.

Inclusion

- Identify genuine partnerships or collaborations between private and voluntary organizations and community members that represent a high level of trust.
- Collaborate with local community educational centers to enhance disaster resilience education courses and through curriculum at all educational levels.
- Engage in cultural/community activities (e.g., festivals, church meetings) to increase the awareness of mutual aid and knowledge of safe locations, and access to emergency facilities and communication.
- Learn community codes and regulations.
- Use special training and support programs for the most vulnerable members of the community (e.g., elderly, disabled).
- Demand representation of various vulnerable groups within the community in the decision-making process.
- Promote a comprehensive awareness campaign of how to manage

disaster risks and access to available services and facilities designed for various groups in the community by official agencies.
- Design emergency response covering physical, financial, and psychological services to protect and help different vulnerable groups within the community.
- Develop open discussions about disaster risks and implications with the community in informal locations such as local coffee shops or elder-care centers to strengthen the participation of the most vulnerable groups in the community.

Guidelines for Resilient Communities

Advocacy

- Create accountable community leadership of the disaster management program, including community vision, set of priorities, and general needs and goals (e.g., poverty alleviation, equality in employment, cultural preservation, quality of life).
- Provide a comprehensive assessment of all disaster risks and relevant socioeconomic, physical, psychological, and cultural implications (both actual and potential) both internal and external to the community, such as public officials, private and local organizations, and volunteers.
- Encourage members of the community to share their views, lessons from the events, problems, and possible solutions.

Competency

- Take the initiative to talk with community members and share with them relevant knowledge regarding their rights and the government's legal and regulatory obligations.
- Develop an ongoing partnership with external agencies and organizations to support the community in times of emergency.
- Educate community members on adequate nutrition, sanitation, hygiene, water, and first aid techniques to provide effective and timely emergency response.
- Increase community activities to support self-confidence and motivate mutual aid and voluntarism.

- Include representatives of all groups and sources of expertise within the community in community decision-making.

Inclusion

- Assimilation of community codes (e.g., warning signs) and agreed regulations when warnings are issued at all levels of community interactions, including family, local labor, and religious groups.
- Develop communication and decision-making processes and terms of coordination between community and other local external organizations.
- Community leadership should try to mobilize volunteers in all levels of resilience planning (not only in first aid).

Notes

Introduction

1. Denhardt and Denhardt, "The New Public Service."
2. Aldrich, *Building Resilience*; Aldrich and Crook, "Strong Civil Society"; Barenstein and Leemann, *Post-Disaster Reconstruction and Change*; Ink, "An analysis"; Kapucu, Rivera, and Hawkins, *Disaster Resiliency*; Menzel, "The Katrina Aftermath"; Nickel and Eikenberry, "Responding to 'Natural' Disasters"; Özerdem and Jacoby, *Disaster Management*; Stivers, "So Poor and So Black."
3. de Silva, "Ethnicity, Politics and Inequality"; Kapucu, "Interagency Communication"; Maxwell et al., "Preventing Corruption"; Stephenson and Schnitzer, "Interorganizational Trust"; Wilder, "Aid and Stability."
4. Ink, "An Analysis"; Koliba, Mills, and Zia, "Accountability"; Waugh, "EMAC, Katrina".
5. Ackerman, *Social Accountability*; Koliba, Mills, and Zia, "Accountability"; Rodan and Hughes, "Ideological Coalitions"; Smith, "The Challenge."

Chapter 1

1. Mileti et al. *Human Systems in Extreme Environments*; Kreps, "Disasters and the Social Order."
2. Kreps and Drabek, "Disasters as Nonroutine Social Problems," 133.
3. Luhmann, *Risk*.
4. IFRC/RCS, *Bam Sends Warning: World Disasters Report 2010*.
5. Cutter et al., "Social Vulnerability"; Quarantelli, *What Is Disaster?*
6. Quarantelli, "Statistical and Conceptual Problems," 332–333.
7. UNISDR. 2004. Living with Risk: A Global Review of Disaster Reduction Initiatives: United Nations.
8. Freudenburg, "Contamination Corrosion, and the Social Order," 26.
9. Alexander, *Natural Disasters*: Quarantelli, *What Is Disaster?*; Quarantelli, "Urban Vulnerability."
10. Tacitus, *Annals*, XV, 40.
11. Tacitus, Cassius Dio, *Roman History*.

12. Griffin, *Nero*.
13. Tacitus, *Annals*, XV, 39.
14. Radice, *The Letters of the Younger Pliny*.
15. Santacroce, "A General Model"; Sheridan et al., "A Model of Plinian Eruptions of Vesuvius"; Sigurdsson et al., "The Eruption of Vesuvius in A.D. 79"; Perret, "Washington, H.S."
16. Canton, "San Francisco 1906 and 2006."
17. Quarantelli, "Statistical and Conceptual Problems."
18. Hansen and Condon, *Denial of Disaster*.
19. Pietz, "Engineering the State."
20. Winchester, *The River at the Center of the World*.
21. National Research Council. *Committee on the Alaska Earthquake*.
22. "Disaster: East Pakistan."
23. Halloran, "Pakistan Storm Relief."
24. Spence, *The Search for Modern China*.
25. U.S. Department of Homeland Security.
26. Haddow et al., *Introduction to Emergency Management*; Neal, "Reconsidering the Phases of Disaster."
27. Waugh, *Living with Hazards*.
28. Clark, "Implementation," 222.
29. Dynes, "Community Emergency Planning."
30. Hanna, "Efficiencies of User Participation"; Sutinen and Kuperan, "A Socio-Economic Theory."
31. Pokorny and Storek, "Current Development"; Sovjakova, "Floods in July 1997 (Czech Republic)."
32. Sovjakova, "Floods in July 1997."
33. Sabatier, "Top-Down and Bottom-Up Models"; Le Grand, *The Other Invisible Hand*.
34. United Nations Development Programme, *Evolution of a Disaster Risk Management System*.
35. Ibid.
36. "Iran Lowers Bam Earthquake Toll."
37. International Federation of Red Cross and Red Crescent Societies, *Bam Sends Warning*. It should be noted that the disaster management efforts in the case of Iran (2003) were a result of successful coordination in a previous disaster that occurred in 1997. This emphasizes the appropriateness of a bottom-up rather than top-down approach to explore the interactive network that functioned in both disasters in Iran.
38. "Words into Action."

Chapter 2

1. Barrows, "Geography as Human Ecology"; White, *Human Adjustment to Floods*.
2. White, *Natural Hazards*; White, *Choice of Adjustment to Floods*; Kates, "Natural Hazard in Human Ecological Perspective"; Boyce, "Let Them Eat Risk?"; Edkins, *Whose Hunger?*; Kates, *Hazard and Choice Perception*; Burton et al., *The Environment as Hazard*.
3. Varley, "The Exceptional and the Everyday"; Alwang et al., "Vulnerability:

Notes 181

   A View from Different Disciplines"; Bankoff, "Rendering the World Unsafe"; Hewitt, *Interpretations of Calamity*.
4. Roberts et al., "Quantification of Vulnerability to Natural Hazards."
5. Hoffman and Oliver-Smith, *Catastrophe and Culture*.
6. Birkmann, "Measuring Vulnerability."
7. Sen, "Capability and Well-Being"; Sen, *Development as Freedom*; Nussbaum, *Women and Human Development*; Nussbaum, "Capabilities as Fundamental Entitlements."
8. Sen, *Commodities and Capabilities*.
9. Sen, *Poverty and Famines*.
10. Saith, "Capabilities: The Concept and Its Operationalisation"; Robeyns, "Sen's Capability Approach"; Stewart and Deneulin, "Amartya Sen's Contribution."
11. Peter, "Political Equality of What?"; Pogge, "Can the Capability Approach Be Justified?"; Deneulin, "Examining Sen's Capability Approach"; Faber and Miller, "Justice and Culture."
12. Chambers, "Editorial Introduction."
13. Ibid., 1.
14. Bohle, *Vulnerability and Criticality*.
15. McCarthy et al., *Climate Change, 2001*; Drimie and Casale, "Multiple Stressors in Southern Africa.
16. Webb and Harinarayan, "A Measure of Uncertainty."
17. Ellis, "Human Vulnerability and Food Insecurity."
18. Devereux, *Identification of Methods and Tools for Emergency Assessments*.
19. Ibid., 1.
20. See similar conceptualization of levels of risk exposure on the population level in CARE, *Managing Risks, Improving Livelihoods*.
21. Devereux, *Identification of Methods and Tools for Emergency Assessments*, 11.
22. Ibid., 8.
23. Watts and Bohle, "The Space of Vulnerability."
24. Bohle, *Vulnerability and Criticality*.
25. Chambers, "Editorial Introduction."
26. Blaikie et al., *At Risk*.
27. Ibid., 11.
28. Riely, "A Comparison of Vulnerability Analysis Methods and Rationale."
29. Ibid., 2.
30. Dilley and Boudreau, "Coming to Terms with Vulnerability," 237.
31. Dilley and Boudreau, "Coming to Terms with Vulnerability."
32. Flanagan et al., "A Social Vulnerability Index for Disaster Management."
33. Kirby, "Theorising Globalisation's Social Impact"; Scholte, *Globalization*.
34. Harriss-White, *Globalization and Insecurity*, 3.
35. Hampson and Hay, "Human Security."
36. Stiglitz, *The Roaring Nineties*, 20.
37. UN, *Report on the World Social Situation*, 14.
38. Ibid.
39. Ibid., 2.
40. Ibid., 15.

41. UNDP, *Human Development Report*, 90.
42. Mishra, *Globalization and the Welfare State*, 70.
43. ECLAC, *Equity, Development and Citizenship*, 52.
44. Ibid.
45. Watson et al., "Climate Change 1995."
46. UNEP, "Global Environment Outlook 3," 302–303.
47. ESCAP, *Economic Vulnerability*.
48. IMF, "Vulnerability Indicators."
49. World Bank, *World Development Report 2000-01*.
50. Briguglio, "Some Conceptual and Methodological Considerations." For further efforts to construct regional vulnerability index, see Naudé et al., "Measuring the Vulnerability of Subnational Regions in South Africa."
51. World Bank, *Globalization, Growth, and Poverty*; Wade, "The Disturbing Rise in Poverty and Inequality"; Kirby, "Theorising Globalisation's Social Impact," 640–641.
52. Cutter et al., "Social Vulnerability."
53. Ibid., 11.
54. ESCAP, "Economic Vulnerability," 25.
55. Kirby, "Theorising Globalisation's Social Impact."
56. Polanyi, *The Great Transformation*, 29.
57. Polanyi, *The Livelihood of Man*.
58. Polanyi, "Our Obsolete Market Mentality," 73.
59. Ibid., 164–165.
60. Kirby, "Theorising Globalisation's Social Impact."
61. O'Brien and Williams, *Global Political Economy*, 32–34.

Chapter 3

1. McEntire et al., "A Comparison of Disaster Paradigms."
2. Ibid.; Geis, "By design"; Trim, "An Integrative Approach."
3. McEntire et al., "A Comparison of Disaster Paradigms," 275.
4. Ibid., 274.
5. Federal Emergency Management Agency (FEMA). http://www.fema.gov/whole-community.
6. Cautilli et al., "Current Behavioral Models of Client and Consultee Resistance."
7. Miller and Rollnick, *Motivational Interviewing*.
8. Freud, *Freud's Psychoanalytic Procedure*, 261–262.
9. Ibid., 222–223.
10. Brems, *Psychotherapy*; Freud, "On the History of Psychodynamic Movement."
11. Wolf, *Treating the Self*.
12. Geis, "By Design," 151.
13. Mooney, "'Problem' Populations, 'Problem' Places"; Özerdem and Jacoby, *Disaster Management and Civil Society*.
14. Oxfam, "Hurricane Dennis Leaves Behind Destruction in Cuba."
15. Thompson and Gaviria, *Weathering the Storm*.
16. Geis, "By Design," 153.

17. Aguirre, "Cuba's Disaster Management Model."
18. Armstrong, "Back to the Future."
19. IUCN, *United Nations Environment Programme*.
20. World Commission on Environment and Development (Brundtland Commission), *Our Common Future*, 43.
21. Beatly, "The Vision of Sustainable Communities," 243.
22. Godschalk et al., *Catastrophic Coastal Storms*.
23. Mileti, *Disasters by Design*.
24. Ibid., 155–207.
25. Tayag et al., "People's Response to Eruption Warning."
26. McEntire, "Sustainability or Invulnerable Development?"
27. Berke, "Natural Hazard Reduction and Sustainable Development," 14–15.
28. Ibid., 14.
29. Luthar and Ciccetti, "The Construct of Resilience."
30. Walsh, "A Family Resilience Framework."
31. Buckle et al., "New Approaches to Assessing Vulnerability and Resilience," 8–9.
32. Luthar and Ciccetti, "The Construct of Resilience"; Rutter, "Resilience Concepts and Findings."
33. Schoon, *Risk and Resilience*.
34. Luthar and Cicchetti, "The Construct of Resilience."
35. Comfort et al., *Designing Resilience*; Aldrich, *Building Resilience*.
36. Ibid.
37. Barenstein and Leemann, *Post-disaster Reconstruction and Change*; Kapucu, Hawkins, and Rivera, *Disaster Resiliency*.
38. Telford, "Learning Lessons from Disaster Recovery."
39. McEntire et al., "A Comparison of Disaster Paradigms," 268–269.
40. McEntire et al., "A Comparison of Disaster Paradigms."
41. Adams and Balfour, *Unmasking Administrative Evil*; Ink, "An Analysis of the House Select Committee"; Menzel, "The Katrina Aftermath"; Mooney, "'Problem' Populations"; Solnit, *A Paradise Built in Hell*; Stivers, "So Poor and so Black"; Ventriss, "Two Critical Issues"; Britton and Clarke, "From Response to Resilience"; McEntire, "Triggering Agents"; Bok, "Government Personnel Policy"; Cooper, "Big Questions"; Stivers, *Governance in Dark Times*; Krause, *Civil Passions*; Boin et al., "The politics of Crisis Management."

Chapter 4

1. Bell, "Communitarianism."
2. Lawton, *Ethical Management for the Public Services*, 69.
3. Kasher, "Professional Ethics," 15–29.
4. Spicer and Terry, "Legitimacy, History, and Logic."
5. Within the public administration community there has been debate over the whether public administrators should be politically neutral or active citizens. See Waldo, *The Enterprise of Public Administration*, which acknowledged the need for a more active role in public debate; Walzer, "Michael Sandel's America."

6. Frederickson, *Ethics and Public Administration*; Marini, "An Introduction."
7. Reich, *The Power of Public Ideas*.
8. Following Nye, "Corruption and Political Development," 419, official corruption is a "behavior which deviates from the formal duties of a public role because of private-regarding (personal, close family, private clique) pecuniary or status gains; or violates rules against the exercise of certain types of private-regarding influence."
9. Williams, "Persons, Character, and Morality."
10. Chandler, "Deontological Dimensions of Administrative Ethics"; Kant, "Fundamental Principles of Metaphysics of Morals"; Pugh, "The Origins of Ethical Frameworks."
11. Frederickson, *Ethics and Public Administration*, 248.
12. Ibid.; Guy, "Using High Reliability Management"; Lewis, *The Ethics Challenge*.
13. Chandler, "Deontological Dimensions."
14. Garfalo and Geuras, *Ethics in Public Administration*; *Practical Ethics in Public Administration*.
15. Svara, *The Ethics Primer for Public Administrators*.
16. *Webster's Third New International Dictionary*, 1131, 1521–1522.
17. Dominick, "Neutral Is Not Impartial."
18. Brighouse and Swift, "Legitimate Parental Partiality"; Cottingham, "Partiality: Favouritism and Morality"; Scheffler, *Boundaries and Allegiances*; Williams, "Persons, Character, and Morality."
19. Foot, "When Is a Principle a Moral Principle?"; Anscombe, "Modern Moral Philosophy."
20. Assessing the intrinsic moral value of partiality requires some sense of what it is that makes close relationships morally right, what it is that determines the moral value they have as relationships. To conceptualize partiality in moral terms we prefer to use Robert Goodin's view of relationships (Goodin, *Protecting the Vulnerable*.). For Goodin, relationships are morally justified to the extent that they are actually successful in promoting the protection of the vulnerable. Thus, partiality is morally required to the extent that it contributes to the protection of those who are vulnerable. Close friendships and relations are especially well placed to offer such protection to each of us because concern and frequency of interaction are typically greater in those relationships than in others (Blum, *Friendship, Altruism, and Morality*; Cottingham, "Partiality: Favouritism and Morality"; Goodin, "What Is So Special"; Oldenquist, "Loyalties."
21. Nussbaum, "The Discernment of Perception."
22. Thomas, "Reasonable Partiality."
23. MacIntyre, *Whose Justice?*; Miller, *On Nationality*; Tamir, *Liberal Nationalism*; Scheffler, *Boundaries and Allegiances*, 83–95, 56–65.
24. Gilligan, *In a Different Voice*.
25. Adams and Balfour, *Unmasking Administrative Evil*; Bok, "Government Personnel Policy"; Cooper, "Big Questions in Administrative Ethics"; Krause, *Civil Passions*; Moe, "The Politicized Presidency"; Stivers, "The Listening Bureaucrat"; *Governance in Dark Times*; Woller and Patterson, "Public Administration Ethics."

26. Brady, "'Publics' Administration"; O'Leary, *The Ethics of Dissent*; Quill, "Ethical Conduct and Public Service."
27. Adams and Balfour, *Unmasking Administrative Evil*; O'Leary, *The Ethics of Dissent*; Stivers, *Governance in Dark Times*.
28. Nagel, *Equality and Partiality*; Scheffler, *The Rejection of Consequentialism*.
29. Stivers, "The Listening Bureaucrat"; Geis, "By Design," 154.
30. Box, *Citizen Governance*; King and Stivers, "Government Is Us"; McSwite, *Legitimacy in Public Administration*; Van Wart, *Changing Public Sector Values*.
31. Zanetti, "Advancing Praxis"; Frazer and Lacey, *The Politics of Community*.
32. Aristotle, *The Politics*, 1253a1.
33. See MacIntyre, *After Virtue*; Sandel, *Liberalism and the Limits of Justice*; Taylor, *Sources of the Self*; Walzer, *Spheres of Justice*.
34. Ackerman, "Why Dialogue?"
35. Other communitarian scholars are Bell, *Communitarianism and Its Critics*; and Glendon, "Rights Talk."
36. For commentators over the debate between liberals and communitarians, see Gardbaum, "Law, Politics, and the Claims of Community"; Neal and Paris, "Liberalism and the Communitarian Critique"; and especially Mulhall and Swift, *Liberals and Communitarians*; Fraser and Lacey, *The Politics of Community*; Okin, *Justice, Gender, and the Family*. For discussions related to the concept of the self in the debate, Beiner, *What's the Matter with Liberalism?*; Carse, *The Liberal Individual*; Elshtain, "The Communitarian Individual.
37. Avineri and de Shalit, *Communitarianism and Individualism*.
38. Taylor, *Philosophical Papers*, 35.
39. According to Etzioni's view, "the admittedly more complex concept of a self congenitally contextuated within a community" ("A Moderate Communitarian Proposal," 157–158.
40. MacIntyre, *Whose Justice?* 98; see also MacIntyre, "Critical Remarks"; "A Partial Response"; "Plain Persons and Moral Philosophy"; "Moral Philosophy"; "The Magic in the Pronoun 'My'"; "Moral Rationality"; "How Moral Agents Became Ghosts".
41. Sandel, *Liberalism and the Limits of Justice*, 150; see also "Moral Argument and Liberal Toleration"; and "Political Liberalism.
42. Taylor, "Justice after Virtue," 38.
43. Taylor, *Philosophical Papers*; "Justice After virtue."
44. For liberal responses to the communitarian criticism, see: Gutmann, "Communitarian Critics of Liberalism"; Kukathas, "Against the Communitarian Republic"; Kymlicka, "Communitarianism, Liberalism, and Superliberalism"; "Liberalism and Communitarianism"; Moore, *Foundations of Liberalism*; "Justice for Our Times"; Buchanan, "Assessing the Communitarian Critique of Liberalism"; Caney, "Liberalisms and Communitarians"; "Liberalism and Communitarianism"; "Rawls, Sandel, and the Self"; Cohen, *The Symbolic Construction of Community*; Conway, *Classical Liberalism*, 60–100; and Paul and Miller, Jr., "Communitarian and Liberal Theories."
45. Walzer, *Spheres of Justice*, 56.

46. Ibid., 31.
47. Ibid.
48. Ibid., 314.
49. MacIntyre, *After Virtue*, 220.
50. See, for example, Taylor, "Hegel: History and Politics," 182; and *Sources of the Self*, 25–52.
51. Sandel, *Liberalism and the Limits of Justice*, 179.
52. Ibid., 150.
53. Kymlicka, *Liberalism, Community and Culture*.
54. Sandel, *Democracy's Discontent*, 13.
55. Ibid., 203.
56. Sandel, "The Procedural Republic," 87.
57. See also Miller's discussion, in "Nature, Justice, and Rights," in Aristotle's Politics, of the "natural priority of the polis" (45–56), particularly the "completeness interpretation" (50–53).
58. Selznick, "The Idea of a Communitarian Morality," 5.
59. MacIntyre, "Moral Philosophy," 207–208.
60. Taylor, "Alternative Futures"; "Cross-Purposes."
61. Kymlicka, *Contemporary Political Philosophy*, 122–123.
62. Etzioni, *The Essential Communitarian Reader*, xii.
63. Bellah, "Community Properly Understood," 16.
64. Selznick, "The Idea of a Communitarian Morality," 7.
65. Ibid., 11.
66. Dworkin, *A Matter of Principle*, 230.
67. Beiner, *What's the Matter with Liberalism?* 14.
68. Etzioni, *The Spirit of Community*, 260.
69. Diamond et al., *Politics in Developing Countries*.
70. Ibid., 27.
71. Stepan, *Rethinking Military Politics*, 3.
72. Huber et al., "The Impact of Economic Development on Democracy."
73. Ibid., 73.
74. Schmitter, "Civil Society and Democratization," 59.
75. Putnam, *Bowling Alone*.
76. See for example, Cohen and Arato, *Civil Society and Political Theory*; Walzer, "The Idea of Civil Society"; and Bratton, "Beyond the State."
77. Keane, *Democracy and Civil Society*, 19.
78. Young, *Inclusion and Democracy*, 156.
79. Friedrich, "Public Policy."
80. Mass and Radway, "Gauging Administrative Responsibility."
81. Stivers, "The Public Agency as Polis"; "The Listening Bureaucrat," 86.
82. Frederickson, "The Recovery of Civism."
83. Crosby et al., "Citizens Panels."
84. de Leon, "The Democratization of the Policy Sciences."

Chapter 5

1. Putnam, "The Prosperous Community," "Bowling Alone."
2. Walzer, "Equality and Civil Society," 47.
3. Streich, "Constructing Multiracial Democracy."

4. Walzer, *Spheres of Justice*, 39.
5. Ibid., 89.
6. Walzer, "Response," 294.
7. Paton et al., "Disaster Response."
8. Walzer, *Spheres of Justice*, 29.
9. Ibid., 31.
10. Ibid., 8.
11. Ibid., 314.
12. Walzer, "Seminar with Michael Walzer," 239.
13. Potter, *How Can I be Trusted?* 7.
14. Fukuyama, *Trust*.
15. Redding, *The Spirit of Chinese Capitalism*, 66.
16. Walzer, *Spheres of Justice*, 92–93.
17. Ibid., xiii.
18. Ibid., xi–xii.
19. Ibid., 10–11.
20. Ibid., 10.
21. Ibid., 20.
22. Ibid., xiv.
23. Ibid., 14.
24. Walzer, "Response," 32.
25. Walzer, *Spheres of Justice*, 29.
26. Ibid., 164.
27. Ibid., 17–20.
28. Ibid., 6.
29. Ibid., 7.
30. Ibid., 8.
31. Ibid.
32. Ibid., 8–9.
33. Ibid., 9.
34. Walzer, "Socializing the Welfare State," 300.
35. Walzer, *Interpretation and Social Criticism*, 39.
36. Action for Advocacy, 2.
37. *Merriam-Webster's Collegiate Dictionary*, 10th ed., 18.
38. Bramlett et al., "Comsumercentric Advocacy."
39. American Nurses Association, *Code of Ethics*.
40. Curtin, "The Nurse as Advocate."
41. Ibid.; Mallik, "Advocacy in Nursing."
42. Lewis et al., "Community Counseling."
43. Lewis et al., *ACA Advocacy Competencies*, 2.
44. Ibid.
45. Cochran, "The Parental Empowerment Process"; Rappaport, "In Praise of Paradox."
46. Walzer, "Response," 282.
47. Walzer, *Spheres of Justice*, 314.
48. Green, *Building Robust Competencies*; Lucia and Lepsinger, *The Art and Science of Competency Models*; Mansfield, "Building Competency Models"; Rodriguez et al. "Developing Competency Models".
49. Boyatzis, *The Competent Manager*, 6.

50. Boyatzis, "Beyond Competence," 9; Spencer and Spencer, *Competence at Work*, 7.
51. McClelland, "Testing for Competence," 2.
52. Vazirani, "Competencies and Competency Model."
53. Walzer, *Spheres of Justice*, xiv.
54. Ibid., 14.
55. Walzer, "Socializing the Welfare State," 300.
56. Guildford, *Making the Case for Social and Economic Inclusion*.
57. Silver, "Reconceptualizing Social Disadvantage," 63.
58. Walker, *Britain Divided*; Bhalla and Lapeyre, "Social Exclusion."
59. Room, *Beyond the Threshold*.
60. Bhalla and Lapeyre, "Social Exclusion."
61. Ibid., 430.
62. Ibid., 418.
63. Walzer, *Interpretation and Social Criticism*, 46.
64. Solnit, *A Paradise Built in Hell*.
65. Walzer, *Spheres of Justice*, 89.
66. Goodwin-Smith, "Something More Substantive."

Chapter 6

1. Ink, "An Analysis of the House Select Committee"; Menzel, "The Katrina Aftermath"; Stivers, "So Poor and So Black"; Adams and Balfour, *Unmasking Administrative Evil*.
2. BondGraham, "The New Orleans that Race Built."
3. Goss, "Civil Society and Civic Engagement."
4. Adams and Balfour, *Unmasking Administrative Evil*; Ink, "An Analysis of the House Select Committee"; Menzel, "The Katrina Aftermath"; Nickel and Eikenberry, "Responding to 'Natural' Disasters"; Stivers, "So Poor and So Black."
5. BondGraham, "The New Orleans that Race Built."
6. http://indahnesia.com/indonesia/event/61/west_sumatra_earthquakes.php.
7. Hsu, "2 million Displaced by Storms."
8. Ministry of Civil Affairs of the People's Republic of China, "The General Review of Earthquake Disaster."
9. Skowronek, *Building a New American State*.
10. It should be noted that the law did not cover state and municipal governments.
11. Hatch Act.
12. U.S. Department of the Interior.
13. On January 1, 1978, the Office of Personnel Management was reestablished under the provisions of Reorganization Plan No. 2 of 1978 (43 F.R. 36037, 92 Stat. 3783) and the Civil Service Reform Act of 1978.
14. U.S. Code Title V.
15. U.S. Department of Labor, "Federal Government, Excluding the Postal Service".
16. Krugman, *The Great Unraveling*.
17. Universal Declaration of Human Rights (UDHR); American Declaration of the Rights and Duties of Man.

18. George W. Bush, "Katrina Address," New Orleans, September 15, 2005. Available at www.foxnews.com/story/0,2933,169514,00.html.
19. Forgette et al., "Race, Hurricane Katrina"; Stivers, "So Poor and So Black."
20. Landphair, "The Forgotten People of New Orleans," 837.
21. NESRI, "A Constant Threat."
22. Krieger and Higgin, "Housing and Health"; Jacobs et al., "The Relationship of Housing and Population Health"; Brunker, "Are FEMA Trailers 'Toxic Tin Cans'?"; Kaplan, "FEMA Covered up Cancer Risks"; Spake, "Dying for a Home"; Manuel, "In Katrina's Wake"; Center for Housing Policy, *The Positive Impacts of Affordable Housing*; Wallace and Wallace, *A Plague on Your Houses*. See also WHO Commission on Social Determinants of Health; and Commission on Social Determinants of Health Knowledge.
23. McAllister et al., "Root Shock Revisited"; Fullilove, "Root Shock."
24. Henry J. Kaiser Family Foundation, *Health Challenges for the People of New Orleans*.
25. NESRI, "A Constant Threat."
26. http://reachnola.org/.
27. Hummel and Ahlers, *Lessons from Katrina*.
28. Duval-Diop et al., "Enhancing Equity with Public Participatory GIS."
29. http://www.maydaynolahousing.org/.
30. NESRI, "International Advisory Group."
31. http://www.nocog.org/New_Orleans_Coalition_On_Open_Governance/Home.html.
32. http://thelensnola.org/.
33. http://www.law.tulane.edu/plc/.
34. http://www.la-par.org/whatispar.cfm.
35. http://www.puentesno.org/about-us.html.
36. Simon, "New School Era Opens Today."
37. Chubb and Moe, "Politics, Markets and the Organization of Schools."
38. Rasheed, "Chaos and Hope in New Orleans Schools," 44.
39. Giroux, *Stormy Weather*; Hill and Hannaway, *After Katrina*; Mooney, "'Problem' Populations, 'Problem' Places"; Ritea, "Public Schools' Makeup Similar," "Skeleton Crew Left to Gut N.O. System," "School District Pledges"; Warner, "Demographer Says Many Residents Want to Return."
40. Tuzzolo and Hewitt, "Rebuilding Inequity."
41. Tyack and Cuban, *Tinkering Towards Utopia*.
42. Atkins, "Restoring Mississippi's Gulf Coast Culture."
43. Cross et al., *Towards a Culturally Competent System of Care*.
44. Nobles, *African Psychology*.
45. Hsu, "2 million Displaced by Storms"; Krause, "New Orleans."
46. www.Census.gov.
47. Hernandez and Isaacs, *Promoting Cultural Competence*.
48. Weil, "The Rise of Community Engagement"; Colten, "Vulnerability and Place"; Elliott and Pais, "Race, Class, and Hurricane Katrina"; Henkel et al., "Institutional Discrimination"; Hartman and Squires, *There Is No Such Thing as a Natural Disaster*; Herring, "Hurricane Katrina"; Chamlee-Wright and Storr, "Club Goods."

49. Weil, "The Rise of Community Engagement After Katrina," 3.
50. http://www.mqvncdc.org/.
51. Jones and Whitney, "Dissolving Barriers," 60.
52. http://safestreetsnola.org/.
53. Cooper and Block, *Disaster*, 172, 257, 266; Solnit, *A Paradise Built in Hell*.
54. Jurkiewicz, "Louisiana's Ethical Culture"; Waugh, *Living with Hazards*.
55. Irons, "Rebuild?"; Terry, "The Thinning of Administrative Institutions."
56. Anderson, "The Idea of Power in Javanese Culture."
57. Emmerson, "The Bureaucracy in Political Context."
58. King, "Civil Service Policies in Indonesia."
59. World Bank, "Pay and Patronage," 33–36.
60. Rehabilitation and Reconstruction Agency, "Aceh and Nias One Year After the Tsunami Report."
61. Leon et al., "Capacity Building Lessons"; Petal et al., "Community-Based Construction."
62. Kennedy et al., "The Meaning of Build Back Better."
63. UNHCR, *Handbook for Emergencies*.
64. Sudmeier-Rieux et al., *Ecosystems*.
65. Government of Indonesia, "Indonesia: Preliminary Damage and Loss Assessment"; "Indonesia: Notes on Reconstruction/Assessment"; "Master Plan for the Rehabilitation and Reconstruction." Hedman, "Back to the Barracks"; "A State of Emergency, a Strategy of War."
66. Ibid.
67. See, for example, "Protesters Attack Aceh Tsunami Reconstruction Office," 11: According to Banda Aceh Deputy Police Chief Dede Setyo, the police "decided to disperse the crowd because they had been staying outside the Rehabilitation and Reconstruction Agency (BRR) offices beyond the timeline that we gave them." In September 2006, BRR cited figures of 70,000 (BRR, 2006b) and 100,000 (BRR, 2006a). BRR, "Demonstrations and Blockade of BRR in Banda Aceh"; "Special Unit on Barracks."
68. Centers for Disease Control and Prevention, "Assessment of Health-Related Needs."
69. Nazara and Resosudarmo, "Aceh-Nias Reconstruction and Rehabilitation," 21; Ross, "Resources and Rebellion."
70. Ross, "Resources and Rebellion"; Moore, "Aceh Education."
71. Hestyanti, "Children Survivors."
72. Rehabilitation and Reconstruction Agency, "Aceh and Nias One Year After the Tsunami."
73. Souza et al., "Mental Health Status of Vulnerable Tsunami-affected Communities."
74. Tsunami Evaluation Coalition, *Impact of the Tsunami Response*, 31.
75. Norris et al., "Community Resilience as a Metaphor"; Paton, *Measuring and Monitoring Resilience*; Cutter et al., "Disaster Resilience Indicators"; Longstaff et al., "Building Resilient Communities"; Shreib et al., "Measuring Capacities for Community Resilience."
76. Schott, "'Smong' Legend Becomes a Lifesaver."

77. Longstaff et al., "Building Resilient Communities."
78. Pelling and High, "Understanding Adaptation."
79. Aldrich, "Fixing Recovery."
80. UNDP, *Civil Society in Aceh*; Bowen, "On the Political Construction of Tradition."
81. Thorburn et al., "The Acehnese Gampong Three Years On."
82. Tsunami Evaluation Coalition, *Impact of the Tsunami Response*, 31.
83. Harper, *Guardianship, Inheritance and Land Law*.
84. Ibrahim, "Kewibawaan Dalam."
85. Sujoti and Rahman, *Membangun Aceh Dari Gampong*; Kenny, "Reconstruction in Aceh"; KDP, *Village Survey in Aceh*.
86. Dasgupta and Beard, "Community Driven Development."
87. Bachyul, "Earthquake Victims."
88. Syarif, *Gampong dan Mukin di Aveh*, 12.
89. Sulaiman, "From Autonomy to Periphery," 125–126.
90. Ibid.
91. Jing, "History and Context."
92. Chien and Li, "Civil Service Law"; Denigan, "Defining Public Administration."
93. National Bureau of Statistics, *China Statistical Yearbook 2003*, 132.
94. Ministry of Personnel, "Provisional Regulations on Civil Service Position Exchange."
95. Liu, *Administrative Reform in China*.
96. Brownlee, *Authoritarianism*.
97. Jacobs and Wong, "China Reports Student Toll for Quake."
98. Human Rights in China, "Family Visits Still Denied."
99. Wong, "China Presses Hush Money on Grieving Parents."
100. Human Rights in China, "Human Rights in China Condemns."
101. Human Rights in China, "Family Visits Still Denied."
102. "China Detains Quake School Critic."
103. Ibid.
104. Eunjung, "School Hit by the Sichuan Earthquake."
105. Jacobs and Wong, "China Reports Student Toll for Quake"; Larmer, "Sichuan Earthquake."
106. "Chinese Government Promises Inquiry into Shoddy Construction."
107. Chan, "The Untold Stories."
108. Xinhua News Agency, "Beijing, Jiangxi, Hunan, Inner Mongolia."
109. Wang et al., *Wenchuan Dizhen Gongmin Xingdong Baogao*.
110. Chen et al. "Long-term Psychological Outcome of 1999 Taiwan Earthquake Survivors"; Wang et al., *Wenchuan Dizhen Gongmin Xingdong Baogao*.
111. Wang et al., *Wenchuan Dizhen Gongmin Xingdong Baogao*.
112. Anastasio et al., "Increased Gender-Based Violence."
113. Chan et al., "The Untold Stories."
114. Bonomi et al., "Health Care Utilization."
115. Chan et al., "Female Victimization."
116. Bermudez et al. "Mental Health of Women Battered by their Partners"; Bonomi et al., "Health Care Utilization"; Kim et al. "The Incidence and

Impact of Family Violence"; Vung et al., "Intimate Partner Violence Against Women."
117. Give2Asia, "Sechuan Earthquake Relief and Recovery."
118. Walzer, *Spheres of Justice*, 75.
119. Zhuang et al., "Research on the Influencing Factors of Peasants' Autonomy."
120. Gu, "The Key to Encourage the Victims."
121. Zhuang et al., "Research on the Influencing Factors of Peasants' Autonomy."
122. Jia, "History and Context"; Yang, "A Civil Society Emerges"; Gadsden, "Earthquake Rocks Civil Society."
123. Schwartz and Shieh, *State and Society Responses to Social Welfare Needs in China*; For studies on how crises can also justify authoritarianism, see; Thornton, "Crisis and Governance"; Jessica Teets, "Post-Earthquake Relief."
124. Pekkanen, "Japan's New Politics"; Özerdem and Jacoby, *Disaster Management*; Kubicek, "The Earthquake, Civil Society, and Political Change."
125. White et al. *In Search of Civil Society*, esp. chapter 1; Watson, "Civil Society"; Alagappa, *Civil Society and Political Change in Asia*.. On the debate over civil society in China, see White et al., *In Search of Civil Society*, chapter 1; Brooks and Frolic, *Civil Society in China*; Shieh and Schwartz, "State and Society Responses."
126. Young, "250 Chinese NGOs."
127. For the debated assessment over civic associations in China, see Wang Ming, "The Development and Present Situation"; and Watson, "Civil Society," 188.
128. A Beijing Normal University survey shows that most NGOs arrived in the first few days. See Wang Ming et al., *Wenchuan Dizhen Gongmin Xingdong Baogao*.
129. Jacobs, "Getting the Measure of Management Competence."
130. http://www.redcross.org.cn/ywzd/.
131. http://give2asia.org/aba-teachers-college.
132. http://give2asia.org/grantees-disaster.
133. http://give2asia.org/grantees-eastasia.
134. http://wn.com/Beichuan_County.
135. http://www.cctf-uk.org/CCTF/ html.
136. http://www.monafoundation.org/China-Foundation.
137. http://www.friendsofnature3.com.
138. http://give2asia.org/grantees-china.
139. http://sclf.cri.cn/.
140. http://npodevelopment.org/en/.
141. http://www.chinacsrmap.org/E_OrgShow.asp?CCMOrg_ID=117.
142. http://give2asia.org/beijing-brookseducationcenter.
143. "China Earthquake Relief & Recovery Interim Report." http://www.asiaing.com/china-earthquake-relief-recovery-interim-report.html.
144. http://asiafoundation.org/resources/pdfs/China.pdf.
145. http://give2asia.org/?p=7817.
146. "China Earthquake Relief & Recovery Interim Report." http://www.asiaing.com/china-earthquake-relief-recovery-interim-report.html.

147. http://www.ngodpc.org/.
148. http://www.cfpa.org.cn.
149. "China Earthquake Relief & Recovery Interim Report." http://www.asiaing.com/china-earthquake-relief-recovery-interim-report.html.
150. Stamperdahl Birger, "Two Years after Sichuan Earthquake, Survivors' Lives Returning to Normal." http://committee100.typepad.com/committee_of_100_newslett/2010/06/two-years-after-sichuan-earthquake-survivors-lives-returning-to-normal-1.html.
151. Yuanchang, "Lessons Modern Chinese Charity Workers Can Take"; Office of the State Council, "Notice on Enhancing Management and Use."
152. Frolic, "State-led Civil Society"; Foster, "Embedded in State Agencies."
153. Muramatsu, *Local Power in the Japanese State*; "Postwar Politics in Japan."
154. Johnson, "Get Kids, Pregnant Women Well Clear."
155. Ito, "Administrative Reform in Japan."
156. Clery, "Current Designs."
157. Makinen, "Japan Steps Up Nuclear Plant Precautions."
158. Declaration of Nuclear Emergency. http://www.kantei.go.jp/foreign/topics/2011/20110311Nuclear_Emergency.pdf.
159. Martyn, "Japan Expands Evacuation Zone"; Kaufmann and Penciakova, "Japan's Triple Disaster," 4.
160. Aven, "On Some Recent Definitions."
161. Johnston, "Get Kids, Pregnant Well Clear."
162. Ibid.
163. Sieg, "Japan Government Losing Public Trust."
164. Ibid.
165. Gorbachev, "Turning Point at Chernobyl."
166. http://www.jafs.or.jp/news/touhoku-jishin-2011.html.
167. http://www.ican.or.jp/.
168. http://www.gnjp.org/index_reports.html.
169. http://www.tpak.org/.
170. http://www.genkiokurou.jp/.171. http://www.childfund.or.jp/other/english.html.
172. http://www.worldvision.jp/.
173. http://www.oisca.org/news/.
174. http://acejapan.org/modules/earthquake/index.html.
175. http://blog.canpan.info/adrajapan/category_28/.
176. http://www.cyr.or.jp/cyrblogs/index_000239.html.
177. http://oxfam.jp/.
178. http://www.joicfp.or.jp/jp/donation/tohoku_earth_quake/.
179. http://www.refugee.or.jp/jar/topics/activity/.
180. http://www.baj-npo.org/2011/04/post-35.html.
181. http://www.icajapan.org.
182. http://ameblo.jp/ivyjimukyokublog/entry-10849866028.html.
183. http://www.habitatjp.org/.
184. http://www.peace-winds.org/jp/news/index.html.
185. http://amda.or.jp/.
186. http://share.or.jp/share/.
187. http://www.jocs.or.jp; http://www.mdm.or.jp/nicocoro/.

188. http://www.knk.or.jp/act/JPN/news.html.
189. http://jenhp.cocolog-nifty.com/emergency/.
190. http://www.aarjapan.gr.jp/activity/report/japan/.
191. http://www.careintjp.org/.
192. http://www.jhp.or.jp/sanka/yomu/shinsai.html.
193. http://www.shaplaneer.org/support/jishin_japan.html.
194. http://www.ngo-jvc.net/index.html.
195. http://sva.or.jp/eru/tohoku/.

Chapter 7

1. Barzilai, *Communities and Law*; Benedict, *Patterns of Culture*; Bourdieu, *Outline of a Theory of Practice*; Cohen, *The Symbolic Construction of Community*.
2. Tomasello, *The Cultural Origins of Human Cognition*.
3. Kobila et al., "Accountability in Governance Networks."
4. Egan, *Skilled Helping Around the World*.
5. Gomez and Wilson, "Political Sophistication."

References

Ackerman, Bruce. "Why Dialogue?" *Journal of Philosophy* 6.1 (1989): 5–22.
Ackerman, John M. *Social Accountability in the Public Sector: A Conceptual Discussion.* Social Development Papers: Participation and Civic Engagement, Paper No. 82. Washington, DC: World Bank, 2005.
Action for Advocacy. *The Advocacy Charter.* London: Action for Advocacy, 2002.
Adams, Guy B., and Danny L. Balfour. *Unmasking Administrative Evil.* Armonk and London: M. E. Sharpe, 2009.
Aguirre, Benigno E. "Cuba's Disaster Management Model: Should It Be Emulated?" *International Journal of Mass Emergencies and Disasters* 23.3 (2006): 55–72.
Alagappa, Muthia, ed. *Civil Society and Political Change in Asia: Expanding and Contracting Democratic Space.* Palo Alto, CA: Stanford University Press, 2004.
Aldrich, Daniel "Fixing Recovery: Social Capital in Post Crisis Resilience." *Journal of Homeland Security* June (2010): 1–9.
———. *Building Resilience: Social Capital in Post-Disaster Recovery.* Chicago: University of Chicago Press, 2012.
———, and Kevin D. Crook. "Strong Civil Society as a Double-Edged Sword." *Political Research Quarterly* 61 (2008): 379–389.
Alexander, David. *Natural Disasters.* London: Chapman and Hall, 1993.
Alwang, Jeffrey, Paul Siegel, and Steen Jorgensen. *Vulnerability: A View from Different Disciplines.* Social Protection Discussion Paper Series. No. 0115. World Bank: Washington, DC, 2001.
American Declaration of the Rights and Duties of Man. O.A.S. Res. XXX, Article 9. Adopted by the Ninth International Conference of American States (1948). http://www1.umn.edu/humanrts/oasinstr/zoas2dec.htm. Last accessed March 2010.
American Nurses Association. *Code of Ethics for Nurses with Interpretive Statement.* Washington, DC: American Nurses Publishing, 2001.
Anastario, Michael, Nadin Shehab, and Lynn Lawry. "Increased Gender-Based Violence Among Women Internally Displaced in Mississippi 2 Years

Post-Hurricane Katrina." *Disaster Medicine and Public Health Preparedness* 3.1 (2009): 18–26.
Anderson, Benedict. "The Idea of Power in Javanese Culture." In *Culture and Politics in Indonesia*, edited by Claire Holt, Benedict Anderson, and James Siegel. Ithaca, NY: Cornell University Press, 1972.
Anscombe, Gertrude Elizabeth M. "Modern Moral Philosophy." Reprinted in *Collected Philosophical Papers*, III. Minneapolis: University of Minnesota Press, (1958)1981.
Aristotle. *The Politics*. New York: Penguin, 1967.
Armstrong, Michael. "Back to the Future: Charting the Course for Project Impact." *Natural Hazards Review* 1.3 (2000): 138–144.
Atkins, Joe. "Restoring Mississippi's Gulf Coast Culture, 'One Year After Katrina.'" *Southern Exposure* special report XXXIV.2 (2006): 1–100, esp. pp. 69–70.
Aven, Terje. "On Some Recent Definitions and Analysis Frameworks for Risk, Vulnerability, and Resilience." *Risk Analysis* 31 (2011): 515–522.
Avineri, Shlomo, and Avner De Shalit. *Communitarianism and Individualism*. New York: Oxford University Press, 1992.
Bachyul, S. Jb. Earthquake Victims Still in Hunts in Indonesia. *Jakarta Post*, 2009b. http://www.thejakartapost.com/30/12/2009.
Bankoff, Gregory. "Rendering the world unsafe: 'vulnerability' as Western discourse." *Disasters* 25.1 (2001): 19–35.
Barenstein, Jennifer E. Duyne, and Esther Leemann (eds.). *Post-Disaster Reconstruction and Change: Communities' Perspectives*. Boca Raton, FL: CRC Press, Taylor & Francis Group, 2012.
Barrows, Harlan H. "Geography as Human Ecology." *Annals of the Association of American Geographers* 13 (1923): 1–14.
Bartelt, David W. "On Resilience: Questions of Validity." In *Educational Resilience in Inner-City America*, edited by Margaret C. Wang and Edmund W. Gordon, 97–108. Hillsdale, NJ: Erlbaum, 1994.
Barzilai, Gad. *Communities and Law: Politics and Cultures of Legal Identities*. Ann Arbor, MI: University of Michigan Press, 2003.
Beatley, Timothy. "The Vision of Sustainable Communities." In *Cooperating with Nature: Confronting Natural Hazards with Land-Use Planning for Sustainable Communities*, edited by Raymond Burby, 233–262. Washington, DC: Joseph Henry Press, 1998.
Beiner, Ronald *What's the Matter with Liberalism?* Berkeley, CA: University of California Press, 1992.
———. "Revising the Self." *Critical Review* 8 (1994): 247–256.
Bell, Daniel. *Communitarianism and Its Critics*. Oxford: Oxford University Press, 1993.
———. "Communitarianism." In *The Stanford encyclopedia of philosophy*, edited by Edward N. Zalta. Stanford, CA: Stanford University, 2004. Available at: http://plato.stanford.edu/entries/communitarianism/ (accessed 20 November 2012).
Bellah, Robert N. "Community Properly Understood: A Defense of 'Democratic Communitarianism.'" In *The Essential Communitarian Reader*,

edited by Amitai Etzioni, 15–20. Lanham, MD: Rowman and Littlefield, 1998.
Benedict, Ruth. *Patterns of Culture*. Boston: Houghton Mifflin Company, 1934.
Berke, Phillip R. *Natural Hazard Reduction and Sustainable Development: A Global Assessment*. Center for Urban and Regional Studies, Working Paper no. S95-02, 1995.
Bermudez, Maria P., M. Pilar Matud, and Gualberto Buela-Casal. "Mental Health of Women Battered by Their Partners in El Salvador." *Revista Mexicana de Psicologia* 26.1 (2009): 51–59.
Bhalla, Ajit, and Frederic Lapeyre. "Social Exclusion: Towards an Analytical and Operational Framework." *Development and Change* 28 (1997): 413–433.
Birkmann, Jörn "Measuring Vulnerability." In *Measuring Vulnerability to Natural Hazards*, edited by J. Birkmann, 9–54. New York: United Nations University Press, 2006.
Blaikie, Piers, Terry Cannon, Ian Davis, and Ben Wisner. *At-Risk: Natural Hazards, People's Vulnerability, and Disasters*. London: Routledge, 1994.
Blum, Lawrence A. *Friendship, Altruism, and Morality*. London: Routledge & Kegan Paul, 1980.
Bohle, Hans-Georg. *Vulnerability and Criticality: Perspectives from Social Geography*. Bonn: International Human Dimensions Programme on Global Environmental Change (IHDP), 2001.
Boin, Arjen, Paul 't Hart, Eric Stern, and Bengt Sundelius. "The Politics of Crisis Management: Public Leadership Under Pressure." *Public Administration Review* 69.1 (2009): 159–161.
Bok, Derek. "Government Personnel Policy in Comparative Perspective." In *For the People: Can We Fix Public Service?* edited by John D. Donahue and Joseph S. Nye Jr., 255–272. Washington, DC: Brookings Institution, 2003.
BondGraham, Darwin. "The New Orleans that Race Built: Racism, Disaster, and Urban Spatial Relationships." *Souls* 9.1 (2007): 4–18.
Bonomi, Amy E., Melissa L. Anderson, Federik P. Rivara, and Robert S. Thompson. "Health Care Utilization and Costs Associated with Physical and Nonphysical-Only Intimate Partner Violence." *Health Services Research* 44.3 (2009): 1052–1067.
Bourdieu, Pierre. *Outline of a Theory of Practice*. Cambridge: Cambridge University Press, 1977.
Bowen, John R. "On the Political Construction of Tradition." *The Journal of Asian Studies* 45.3 (1986): 545–561.
Box, Richard C. *Citizen Governance: Leading American Communities into the 21st Century*. Thousand Oaks, CA: Sage Publications, 1998.
Boyatzis, Richard E. *The Competent Manager. A Model for Effective Performance*. New York: Wiley, 1982.
———. "Beyond Competence: The Choice to Be a Leader." *Human Resource Management Review* 3.1 (1993): 1–14.
Boyce, James "Let Them Eat Risk? Wealth, Rights and Disaster Vulnerability." *Disasters* 24.3 (2000): 254–261.

Brady, F. Neil "'Public' Administration and the Ethics of Particularity." *Public Administration Review* 63.5 (2003): 525–534.

Bramlett, Martha H., Sarah H. Gueldner, and Richard L. Sowell. "Consumercentric Advocacy: Its Connection to Nursing Frameworks." *Nursing Science Quarterly* 3.4 (1989): 156–161.

Bratton, Michael. "Beyond the State: Civil Society and Associational Life in Africa." *World Politics* 41 (1989): 407–430.

Brems, Christiane. *Psychotherapy: Process and Techniques.* Boston: Allyn & Bacon, 1999.

Brighouse, Harry, and Adam Swift. "Legitimate Parental Partiality." *Philosophy and Public Affairs* 37.1 (2009): 43–80.

Briguglio, Pascal Lino. "Some Conceptual and Methodological Considerations Relating to the Construction of an Index of Social Vulnerability with Special Reference to Small Island Developing States." In *ECLAC, Towards a Social vulnerability Index in the Caribbean*, 39–62. Spain: ECLAC, 2003.

Britton, Neil R., and Gerard J. Clarke. "From Response to Resilience: Emergency Management Reform in New Zealand." *Natural Hazards Review* 1.3 (2000): 145–150.

Brooks, Timothy, and B. Michael Frolic. *Civil Society in China.* Armonk: M. E. Sharpe, 1997.

Brownlee, Jason. 2007. *Authoritarianism in an Age of Democratization.* Cambridge: Cambridge University Press.

BRR (Rehabilitation and Reconstruction Agency) Indonesia. "Aceh and Nias One Year After the Tsunami Report." Executive Summary December 2005. www.e-aceh-nias.org.

———. Demonstrations and Blockade of BRR in Banda Aceh, BRR International Update, September 20, 2006.

———. Special Unit on Barracks, BRR International Update, September 2006.

Brunker Mike, "Are FEMA Trailers 'Toxic Tin Cans'?" MSNBC, July 25, 2006. http://www.msnbc.msn.com/id/14011193/.

Buchanan, Allen. "Assessing the Communitarian Critique of Liberalism." *Ethics* 99 (1989): 852–882.

Buckle, Philip, Graham Mars, and Syd Smale. "New Approaches to Assessing Vulnerability and Resilience." *Australian Journal of Emergency Management* 15.2 (2000): 8–14.

Buckley, Chris. "China Detains Quake School Critic—Rights Group." *Reuters*, June 17, 2008.

Burton, Ian, Robert W. Kates, and Gilbert F. White. *The Environment as Hazard.* New York: Oxford University Press, 1978.

Caney, Simon. "Rawls, Sandel, and the Self." *International Journal of Moral and Social Studies* 6 (1991): 161–171.

———. "Liberalism and Communitarianism: A Misconceived Debate." *Political Studies* 40 (1992): 273–289.

———. "Liberalisms and Communitarians: A Reply." *Political Studies* 41 (1993): 657–660.

Canton, Lucien G. "San Francisco 1906 and 2006: An Emergency Management Perspective." *Earthquake Spectra* 22 (2006): 159–182.

CARE. *Managing Risks, Improving Livelihoods: Programme Guidelines for Conditions of Chronic Vulnerability*, 2nd ed. Nairobi, Kenya: East and Central Africa Regional Management Unit, 2003.

Carens, Joseph. "A Contextual Approach to Political Theory." *Ethical Theory and Moral Practice* 7 (1989): 117–132.

Carse, Alisa L. "The Liberal Individual: A Metaphysical or Moral Embarrassment?" *Noûs* 28 (1994): 184–209.

Cassius Dio. *Roman History*, Books 62 (c. 229 CE), Vol. VIII of the Loeb Classical Library edition. Cambridge: Harvard University Press, 1925.

Cautilli, Joe D., T. Chris Riley-Tillman, Saul Axelrod, and Phil N. Hineline. "Current Behavioral Models of Client and Consultee Resistance: A Critical Review." *International Journal of Behavioral Consultation and Therapy* 1.2 (2005): 147–54.

www.Census.gov

Center for Housing Policy. *The Positive Impacts of Affordable Housing on Health: A Research Summary*. Washington, DC: Center for Housing Policy and Enterprise Community Partners, 2007. http://www.nhc.org/pdf/chp_int_summary_hsghlth0707.pdf. Last accessed May 2009.

Centers for Disease Control and Prevention (CDC). "Assessment of Health-Related Needs After Tsunami and Earthquake—Three Districts, Aceh Province, Indonesia, July–August 2005." *Morbidity and Mortality Weekly Report* 55 (2006): 93–97

Chambers, Robert. "Editorial introduction: Vulnerability, coping, and policy." *IDS Bulletin* 2.2 (1989): 1–7.

Chamlee-Wright, Emily, and Virgil Henry Storr. "Club Goods and Post-Disaster Community Return." *Rationality and Society* 21.4 (2009): 429–458.

Chan, E. K. Ling, and Yulian L. Zhang. "Female Victimization and Intimate Partner Violence After the May 12, 2008, Sichuan Earthquake." *Violence & Victims* 26.3 (2011): 1–14.

Chan, Emily Y. Y. "The Untold Stories of the Sichuan Earthquake." *The Lancet* 372.9636 (2008): 359–363.

Chandler, Ralph C. "Deontological Dimensions of Administrative Ethics." In *Handbook of Administrative Ethics*, edited by Terry L. Cooper, 147–56. New York: Marcel Dekker, 1994.

Chen, Chin-Hung, Happy Kuy-Lok Tan, Long-Ren Liao, Hsui-Hsi Chen, Chang-Chuan Chan, Joseph-Jror-Serk Cheng, Chung-Ying Chen, Tsu-Nai Wang, and Mong-Liang Lu. "Long-Term Psychological Outcome of 1999 Taiwan Earthquake Survivors: A Survey of a High Risk Sample with Property Damage." *Comprehensive Psychiatry* 48.3 (2007): 269–275.

Chien, Hon S., and Suizhou Li. "Civil Service Law in the People's Republic of China: A Return to Cadre Personnel Management." *Public Administration Review* 67.3 (2007): 383–398.

"China Detains Quake School Critic—Rights Group." *Reuters*, June 17, 2008.

"China Suppresses Angry Parents of Quake-Killed Children." *Tibetan Review: The Monthly Magazine on all Aspects of Tibet* 43.7 (2008): 34–35.

"Chinese Government Promises Inquiry into Shoddy Construction." *World Today* via abc.net.au, May 23, 2008.

Chubb, John E., and Terry M. Moe. "Politics, Markets, and the Organization of Schools." *American Political Science Review* 82.4 (1988): 1065–1087.

Clark, Michael. "Implementation." In *Power and policy in liberal democracies*, edited by Martin Harrap. Cambridge, UK: Cambridge University Press, 1992.

Clery, Daniel. "Current Designs Address Safety Problems in Fukushima Reactors." *Science* 331 (2011): 1506.

Cochran, Moncrieff. "The Parental Empowerment Process: Building on Family Strengths." In *Child Psychology in Action,* edited by John Harris, 12–33. Cambridge, MA: Brookline Books, 1986.

Cohen, Andrew J. "A Defense of Strong Voluntarism." *American Philosophical Quarterly* 35 (1998): 251–265.

Cohen, Anthony P. *The Symbolic Construction of Community.* New York: Routledge, 1985.

Cohen. Jean L., and Andrew Arato. *Civil Society and Political Theory.* Cambridge, MA: MIT Press, 1997.

Colten, Craig E. "Vulnerability and Place: Flat Land and Uneven Risk in New Orleans." *American Anthropologist* 108 (2006): 731–734.

Comfort, Louise K, Arjen Boin, and Chris C. Demchak (eds.). *Designing Resilience: Preparing for Extreme Events.* Pittsburgh, PA: University of Pittsburgh Press, 2010.

Conway, David. *Classical Liberalism: The Unvanquished Ideal.* New York: St. Martin's Press, 1995.

Cooper, Christopher, and Robert Block. *Disaster: Hurricane Katrina and the Failure of Homeland Security.* New York: Henry Holt, 2006.

Cooper, Terry L. "Big Questions in Administrative Ethics: A Need for Focused, Collaborative Effort." *Public Administration Review* 64.4 (2004): 395–407.

Cottingham, John. "Partiality: Favouritism and morality." *Philosophical Quarterly* 36.144 (1986): 357–373.

Crosby, Ned, Janet Kelly, and Paul Schaefer. "Citizens Panels: A New Approach to Citizen Participation." *Public Administration Review* 46 (1986): 170–178.

Cross, Terry, Barbara Bazron, Karl Dennis, and Mareasa Isaacs. *Towards a Culturally Competent System of Care,* vol. I. Washington, DC: Georgetown University Child Development Center, CASSP Technical Assistance Center, 1989.

Curtin, Leah L. "The Nurse as Advocate: A Philosophical Foundation for Nursing." *Advances in Nursing Science* 1.3 (1979): 1–10.

Cutter, Susan L., Bryan J. Boruff, and W. Lynn Shirley. "Social Vulnerability to Environmental Hazards." *Social Science Quarterly* 84.2 (2003): 242–261.

———, Christopher Burton, and Christopher T. Emrich. "Disaster Resilience Indicators for Benchmarking Baseline Conditions." *Journal of Homeland Security and Emergency Management* 7.1 (2010): Article 51.

Dasgupta, Aniruddha, and Victoria A. Beard. "Community Driven Development, Collective Action and Elite Capture in Indonesia." *Development and Change* 38.2 (2007): 229–249.

de Leon, Peter. "The Democratization of the Policy Sciences." *Public Administration Review* 52 (1992): 125–129.
De Silva, Amarasiri M. W. "Ethnicity, Politics and Inequality: Post-Tsunami Humanitarian Aid Delivery in Ampara District, Sri Lanka." *Disasters* 33.2 (2009): 253–273.
Deneulin, Séverine. "Examining Sen's Capability Approach to Development as Guiding Theory for Development policy." DPhil. dissertation, University of Oxford, 2003.
Denhardt, Robert B., and Janet V. Denhardt. "The New Public Service, Serving Rather than Steering." *Public Administration Review* 60.6 (2000): 549–559.
Denigan, Mary. "Defining Public Administration in the People's Republic of China: A Platform for Future Discourse." *Public Performance & Management Review* 24.3 (2001): 215–232.
Devereux, Stephen. *Identification of Methods and Tools for Emergency Assessments to Distinguish Between Chronic and Transitory Food Insecurity, and to Evaluate the Effects of the Various Types and Combinations of Shocks on These Different Livelihood Groups*. Rome: United Nations World Food Programme, 2006.
Diamond, Larry Jay, Juan José Linz, and Seymour Martin Lipset. *Politics in Developing Countries: Comparing Experiences with Democracy*. Boulder, CO: Lynne Rienner Publications, 1995.
Dilley, Maxx, and Tania Boudreau. "Coming to Terms with Vulnerability: A Critique of the Food Security Definition." *Food Policy* 26.3 (2001): 229–247.
"Disaster; East Pakistan: Cyclone May Be the Worst Catastrophe of Century." *New York Times*, November 22, 1970, 169.
Dominick, Donald. "Neutral Is Not Impartial: The Confusing Legacy of Traditional Peace Operations Thinking." *Armed Forces & Society* 29.3 (2003): 415–448.
Drimie, Scott, and Marisa Casale. "Multiple Stressors in Southern Africa: The Link Between HIV/AIDS, Food Insecurity, Poverty and Children's Vulnerability Now and in the Future." *Aids Care* 21.S1 (2009): 28–33.
Duval-Diop, Dominique, Andrew Curtis, and Annie Clark. "Enhancing Equity with Public Participatory GIS in Hurricane Rebuilding: Faith Based Organizations, Community Mapping, and Policy Advocacy." *Community Development* 41.1 (2010): 32–49.
Dworkin, Ronald. M. *A Matter of Principle*. Cambridge, MA: Harvard University Press, 1985.
Dynes, Russell R. "Community Emergency Planning: False Assumptions and Inappropriate Analogies." *International Journal of Mass Emergencies and Disasters* 12 (1994): 141–158.
Economic Commission for Latin America and the Caribbean (ECLAC). *Equity, Development and Citizenship* (LC/G.2071(SES.28/3)). Santiago, Chile: ECLAC, 2000.
Edkins, Jenny. *Whose Hunger? Concepts of Famine, Practices of Aid*. Minneapolis, MN: University of Minnesota Press, 2000.
Egan, Gerald. *Skilled Helping Around the World: Addressing Diversity and Multiculturalism*. Belmont, CA: Brooks/Cole, Cengage Learning, 2006.

Elliott, James R., and Jeremy Pais. "Race, Class, and Hurricane Katrina." *Social Science Research* 35 (2006): 295–321.

Ellis, Frank. "Human Vulnerability and Food Insecurity: Policy Implications." Theme paper produced for the Forum for Food Security in Southern Africa. London: Overseas Development Institute (2003). www.odi.org.uk/food-securityforum.

Elshtain, Jean B. "The Communitarian Individual." In *New Communitarian Thinking: Persons, Virtues, Institutions, and Communities*, edited by Amitai Etzioni, 99–109. Charlottesville, VA: University of Virginia Press, 1996.

Emmerson, Donald K. "The Bureaucracy in Political Context: Weakness in Strength." In *Political Power and Communications in Indonesia*, edited by Karl Jackson and Lucien W. Pye, 82–136. Berkeley: University of California Press, 1978.

ESCAP (UN Economic and Social Commission for Asia and the Pacific). *Economic Vulnerability*. Bangkok, Thailand: ESCAP, 2003.

Etzioni, Amitai. *The Spirit of Community*. New York: Crown Books, 1993.

———. "A Moderate Communitarian Proposal." *Political Theory* 24 (1996): 155–171.

———, ed. *The Essential Communitarian Reader*. Lanham, MD: Rowman and Littlefield, 1998.

Eunjung Cha, Ariana. "School Hit by the Sichuan Earthquake in 2008." *Washington Post* March 28, 2009.

Faber, Cécile, and David Miller. "Justice and Culture: Rawls, Sen, Nussbaum and O'Neill." *Political Studies Review* 1 (2003): 4–17.

Federal Emergency Management Agency (FEMA) in the United States. http://www.fema.gov/whole-community.

Flanagan, Barry E., Edward W. Gregory, Elaine J. Hallisey, Janet L. Heitgerd, and Brian Lewis. "A Social Vulnerability Index for Disaster Management." *Journal of Homeland Security and Emergency Management* 8.1 (2011): Article 3.

Foot, Philippa. "When Is a Principle a Moral Principle?" *Proceedings of the Aristotelian Society* Supp. 28 (1954): 95–109.

Forgette, Richard, Marvin King, and Bryan Dettrey. "Race, Hurricane Katrina, and Government Satisfaction: Examining the Role of Race in Assessing Blame." *Publius: The Journal of Federalism* 38.4 (2008): 671–691.

Foster, Kenneth. "Embedded in State Agencies: Business Associations in Yantai." *China Journal* 47 (2002): 41–65.

Frazer, Elizabeth, and Nicole Lacey. *The Politics of Community: A Feminist Critique of the Liberal-Communitarian Debate*. Toronto: University of Toronto Press, 1993.

Frederickson, H. George. "The Recovery of Civism." *Public Administration Review* 42 (1982): 501–508.

———. *Ethics and Public Administration*. Armonk, NY: M. E. Sharpe, 1993.

Freud, Sigmund. "On the History of Psychodynamic Movement." In *The Standard Edition of the Complete Psychological Works of Sigmund Freud*, vol. 14, edited and translated by James Strachey. London: Hogarth Press, (1914) 1957.

———. *Freud's Psychoanalytic Procedure*. Standard Edition. London: Hogarth Press, 1959.
Freudenburg, William R. "Contamination Corrosion, and the Social Order: An Overview." *Current Sociology* 45 (1997): 19–40.
Friedrich, Carl J. "Public Policy and the Nature of Administrative Responsibility." In *Public Policy: A Yearbook of the Graduate School of Public Administration*, edited by Carl J. Friedrich and Edward S. Mason. Cambridge, MA: Harvard University Press, 1940.
Frolic, B. Michael. "State-Led Civil Society." In *Civil Society in China*, edited by Timothy Brook and B. Michael Frolic, 46–67. Armonk, NY: M. E. Sharpe, 1997.
Fukuyama, Francis. *Trust: The Social Virtues and the Creation of Prosperity*. New York: Penguin, 1995.
Fullilove, Mindy T. "Root Shock: The Consequences of African American Dispossession." *Journal of Urban Health* 78.1 (2001): 72–80.
———. *Root Shock: How Tearing Up City Neighborhoods Hurts America and What We Can Do About It*. New York: Ballantine, 2004.
Gadsden, Amy. "Earthquake Rocks Civil Society." *Far Eastern Economic Review* 171.5 (2008): 25–29.
Gardbaum, Stephen. "Law, Politics, and the Claims of Community." *Michigan Law Review* 90 (1992): 685–760.
Garfalo, Charles, and Dean Geuras. *Ethics in Public Administration: The Moral Mind at Work*. Washington, DC: Georgetown University Press, 1999.
———. *Practical Ethics in Public Administration*. Vienne, VA: Management Concepts, 2005.
Geis, Donald E. "By Design: The Disaster Resistant and Quality-of-Life Community." *Natural Hazards Review* 1.3 (2000): 151–160.
Gilligan, Carol. *In a Different Voice*. Cambridge, MA: Harvard University Press, 1982.
Giroux, Henry A. *Stormy Weather: Katrina and the Politics of Disposability*. Boulder, CO: Paradigm Publishers, 2006.
Give2Asia. "Sichuan Earthquake Relief and Recovery Following the May 12, 2008 Earthquake." Interim Report, May 2009.
Glendon, Mary A. "Rights Talk: The Impoverishment." *Journal of Ethics* 4 (2000): 283–305.
Godschalk, David R., David J. Brower, and Timothy Beatley. *Catastrophic Coastal Storms: Hazard Mitigation and Development Management*. Durham, NC, and London: Duke University Press, 1989.
Gomez, Brad T., and J. Matthew Wilson. "Political Sophistication and Attributions of Blame in the Wake of Hurricane Katrina." *Publius: The Journal of Federalism* 38.4 (2008):633–650.
Goodin, Robert *Protecting the Vulnerable*. Chicago: University of Chicago Press, 1985.
Goodin, Robert E. "What Is So Special About Our Fellow Countrymen?" *Ethics* 98.4 (1988): 663–686.
Goodwin-Smith, Ian. "Something More Substantive than Social Inclusion." *Social Alternatives* 28.2 (2009): 53–37.
Gorbachev, Mikhail. "Turning Point at Chernobyl." April 14, 2006. http://

www.project-syndicate.org/commentary/gorbachev3/English. Last accessed May 2009.
Gove, Philip B., ed. *Webster's Third New International Dictionary.* Springfield, MA: G. & C. Merriam, 1961.

Government of Indonesia (Bappenas). *Indonesia: Preliminary Damage and Loss Assessment. The December 26, 2004 Natural Disaster.* Jakarta: Consultative Group on Indonesia, January 19–20, 2005.
———. *Indonesia: Notes on Reconstruction/Assessment. The December 26, 2004 Natural Disaster.* Jakarta: Consultative Group on Indonesia, January 19–20, 2005.
———. *Master Plan for the Rehabilitation and Reconstruction of the Regions and Communities of the Province of Nanggroe Aceh Darussalam and the Islands of Nias, Province of North Sumatra.* Jakarta: Government of Indonesia, April 12, 2005.
Graham, Room. "Poverty in Europe: Competing Paradigms of Analysis." *Policy and Politics* 23 (1995): 103–113.
Green, Paul C. *Building Robust Competencies: Linking Human Resource Systems to Organizational Strategies.* San Francisco, CA: Jossey-Bass, 1999.
Griffin, Miriam T. *Nero: The End of a Dynasty.* London: Routledge, 2000.
Gu, Shengzu. "The Key to Encourage the Victims in Rebuilding Is Self-Reliance." *The Contemporary Economy* 6 (2008): 1.
Guildford, Janet. *Making the Case for Social and Economic Inclusion.* Halifax, Nova Scotia: Atlantic Region, Health Canada, 2000.
Gutmann, Amy. "Communitarian Critics of Liberalism." *Philosophy and Public Affairs* 14 (1985): 308–322.
Guy, Mary E. "Using High Reliability Management to Promote Ethical Decision Making." In *Ethical frontiers in public management,* edited by James S. Boweman, 185–204. San Francisco: Jossey-Bass Publishers, 1991.
Haddow, George, Jane A. Bullock, and Damon Coppola. *Introduction to Emergency Management*, 3rd Ed. Burlington, MA: Butterworth-Heinemann, 2008.
Halloran, Richard. "Pakistan Storm Relief a Vast Problem." *New York Times*, November 29, 1970, 11–29.
Hampson, Fen Osier, and John B. Hay. "Human Security: A Review of the Scholarly Literature." *The Human Security Bulletin* 1.2 (2002): 1–36.
Hanna, Susan. "Efficiencies of User Participation in Natural Resources Management." In *Property Rights and the Environment: Social and Ecological Issues.* Washington, DC: Beijer Internal Institute of Ecological Economics and the World Bank, 1995.
Hansen, Gladys, and Emmet Condon. *Denial of Disaster: The Untold Story and Photographs of the San Francisco Earthquake of 1906.* San Francisco, CA: Cameron and Company, 1989.
Harper, Erica. *Guardianship, Inheritance and Land Law in Post-Tsunami Aceh.* Banda Aceh: International Development Law Organisation (IDLO), 2006. http://www.idlo.int/publications/19.pdf.
Harriss-White, Barbara, ed. *Globalization and Insecurity: Political, Economic and Physical Challenges.* Basingstoke: Palgrave, 2002.

Hartman, Chester, and Gregory D. Squires, eds. *There Is No Such Thing as a Natural Disaster: Race, Class and Hurricane Katrina.* New York: Routledge, 2006.
Hatch Act. "Political Activity." May 16, 2011. http://www.osc.gov/hatchact.htm.
Hedman, Eva-Lotta E. "Back to the Barracks: Relokasi Pengungsi in Post-Tsunami Aceh, Indonesia." *Indonesia Journal* 80 (2005): 1–19.
———. "A State of Emergency, a Strategy of War: Internal Displacement, Forced Relocation, and Involuntary Return in Aceh." In *Aceh Under Martial Law: Conflict, Violence and Displacement*, RSC Working Paper No. 24. Oxford: University of Oxford, July 2005.
Henkel, Kristin E., John F. Dovidio, and Samuel L. Gaertner. "Institutional Discrimination, Individual Racism, and Hurricane Katrina." *Analyses of Social Issues and Public Policy* 6 (2006): 99–124.
Henry J. Kaiser Family Foundation. *Health Challenges for the People of New Orleans: The Kaiser Post-Katrina Baseline Survey.* Menlo Park, CA/Washington, DC: Kaiser Family Foundation, 2007. http://www.kff.org/kaiserpolls/upload/7659.pdf, p. 15. Last accessed March 2010.
Hernandez, Mario, and Mareasa Isaacs. *Promoting Cultural Competence in Children's Mental Health Services.* Baltimore, MD: Paul H. Brookes Publishing, 1998.
Herring, Cedric. "Hurricane Katrina and the Racial Gulf." *Du Bois Review: Social Science Research on Race* 3 (2006): 129–144.
Hestyanti, Yohana Ratrin. "Children Survivors of the 2004 Tsunami in Aceh, Indonesia. A Study of Resiliency." *Annals of the New York Academy of Science* 1094 (2006): 303–307.
Hewitt, Kenneth, ed. *Interpretations of Calamity: From the Viewpoint of Human Ecology.* Boston MA: Allen and Unwin, 1983.
Hill, Paul, and Jane Hannaway. *After Katrina: Rebuilding Opportunity and Equity into the New New Orleans.* Washington, DC: Urban Institute, 2006.
Hoffman, Susanna, and Anthony Oliver-Smith, eds. *Catastrophe and Culture: The Anthropology of Disaster.* Santa Fe, NM: School of American Research Press, 2002.
Hsu, Spencer. "2 Million Displaced by Storms." *Washington Post*, January 13, 2006: A03.
Huber, Evelyne, Dietrich Rueschmeyer, and John D. Stephens. "The Impact of Economic Development on Democracy." *Journal of Economic Perspectives* 7.3 (1993): 71–85.
Human Rights in China. Press Release: "Human Rights in China Condemns the Detention of Huang Qi by police in Chengdu." June 14, 2008. http://www.hrichina.org/public/contents/56408. Last accessed May 2009.
———. Press Release: "Family Visits Still Denied to Sichuan School Teacher Punished After Quake-Zone Visit." July 29, 2008. http://www.hrichina.org/public/contents/66524. Last accessed June 2009.
Hummel, Rebecca, and Douglas Ahlers. *Lessons from Katrina. The Broadmoor Guide for Planning and Implementation.* Cambridge, MA: Belfer Center for Science and International Affairs. John Kennedy School of Government, Harvard, 2007.

Ibrahim, Syafei. "Kewibawaan dalam Pandangan Masyarakat Aceh" (Authority in the Viewpoint of the Acehnese). *Journal Ilmiah Administrasi Publik* VI.1 (2006). http://publik.brawijaya.ac.id/?hlm=jedlist&ed=1125507600&edid=1135589533. Last accessed March 2010.

ICESCR (International Covenant on Economic, Social and Cultural Rights). G. A. Res. 2200A (XXI), Art. 11, Sec. 1 (1966). http://www2.ohchr.org/english/law/cescr.htm. Last accessed May 2009.

IFRC/RCS (International Federation of Red Cross & Red Crescent Societies). *Bam Sends Warning to Reduce Future Earthquake Risks*, Chapter 4. World Disasters Report 2004: Focus on Community Resilience, Geneva: IFRC, 2004.

IFRC/RCS (International Federation of Red Cross & Red Crescent Societies). *World Disasters Report 2010*. Geneva: Red Cross, 2010.

IMF. "Vulnerability Indicators: A Factsheet." May 2009. http://www.imf.org/external/np/exr/facts/vul.htm. Last accessed July 2011.

Ink, Dwight. "An Analysis of the House Select Committee and White House Reports on Hurricane Katrina." *Public Administration Review* 66.6 (2006): 800–807.

"Iran Lowers Bam Earthquake Toll." *BBC News*, March 29, 2004. http://news.bbc.co.uk/2/hi/middle_east/3579173.stm. Last accessed May 2009.

Irons, John S. "Rebuild? Or Give Money to Residents? Yes." Washington, DC: Center for American Progress, October 6, 2006. http://www.americanprogress.org/kf/katrina%20rebuild%20assistance.pdf.

Ito, Mitsutoshi. "Administrative Reform in Japan: Semi-Autonomous Bureaucracy Under the Pressure Toward a Small Government." In *State and Administration in Japan and Germany*, edited by Michio Muramatsu and Frieder Naschold, 63–78. New York: Walter de Gruyter, 1997.

IUCN (International Union for Conservation of Nature and Natural Resources). *United Nations Environment Programme, World Wildlife Fund, Food and Agriculture Organization of the United Nations*. Paris: UNESCO, 1980.

Jacobs, Andrew, and Edward Wong. "China Reports Student Toll for Quake," *New York Times*, May 7, 2008. http://www.nytimes.com/2009/05/08/world/asia/08china.html?. Last accessed May 2009.

Jacobs, David E., Jonathan Wilson, Sherry L. Dixon, Janet Smith, and Anne Evens. "The Relationship of Housing and Population Health: A 30-Year Retrospective Analysis." *Environmental Health Perspectives* 117.4 (2009): 597–604.

Jacobs, Robin. "Getting the Measure of Management Competence." *Personnel Management* 21.6 (1989): 32–37.

Jia, Xijin. "Chinese Civil Society After the May 12 Earthquake." English translation on Policy Forum Online 08-056A: July 22, 2008. www.nautilus.org/fora/security/08056Jia.html. Last accessed May 2009.

Jing, Yijia. "History and Context of Chinese Public Administration." In *Handbook of Public Administration in East Asia*. London: Taylor & Francis LLC, 2009.

Johnson, Chalmers. *MITI and the Japanese Miracle*. Palo Alto, CA: Stanford University Press, 1982.

Johnston, Eric. "Get Kids, Pregnant Women Well Clear of Nuke Zone: Politicians." *Japan Times*, March 26, 2011. http://search.japantimes.co.jp/cgi-bin/nn20110326b3.html. Last accessed August 2012.

Jones, Catherine, and Jennifer Whitney. "Dissolving Barriers: New Orleans' Latino Health Outreach Project." In *One Year After Katrina, Southern Exposure* special report XXXIV.2 (2006): 1–100.

Jurkiewicz, Carole L. "Louisiana's Ethical Culture and Its Effect on the Administrative Failures Following Katrina." Special issue, *Public Administration Review* 67 (2007): 57–63.

Kant, Immanuel. *Fundamental Principles of Metaphysics of Morals*, translated by Thomas Kingsmill Abbot. Amherst, NY: Prometheus Books, (1785) 1988.

Kaplan, Sheila. "FEMA Covered Up Cancer Risks to Katrina Victims." Salon.com, January 29, 2008. http://www.salon.com/news/feature/2008/01/29/fema_coverup/index.html.

Kapucu, Naim. "Interagency Communication Networks During Emergencies." *American Review of Public Administration* 36.2 (2006): 207–225.

Kapucu, Naim, Fernando Rivera, and Christopher Hawkins, eds. *Disaster Resiliency: Interdisciplinary Perspectives*. New York: Routledge, 2012.

Kasher, Asa. "Professional Ethics." In *Ethical Issues in Mental Health Consultation and Therapy* [*Sogiot Etiot Bmikzohot Hayehotz Vehatipul Hanafshi*], edited by Gabi Schafler, Yehudit Achmon, and Gabriel Weil, 15–29. Jerusalem: Magnes, 2003.

Kates, Robert W. *Hazard and Choice Perception in Flood Plain Management*. Research Paper No. 78. Chicago: Department of Geography, University of Chicago, 1962.

———. "Natural Hazard in Human Ecological Perspective: Hypotheses and Models." *Economic Geography* 47 (1971): 438–451.

Kaufmann, Daniel, and Veronika Penciakova. "Japan's Triple Disaster: Governance and the Earthquake, Tsunami and Nuclear Crises." Brookings Institution, March 17, 2011. http://www.brookings.edu/opinions/2011/0316_japan_disaster_kaufmann.aspx. Last accessed August 2012.

KDP. *2006 Village Survey in Aceh: An Assessment of Village Infrastructure and Social Conditions*. Banda Aceh: The World Bank, 2007. http://siteresources.worldbank.org/INTINDONESIA/Resources/226271-1168333550999/AcehVillageSurvey06_final.pdf. Last accessed March 2010.

Keane, John. *Democracy and Civil Society*. London: Verso, 1988.

Kennedy, Jim, Joseph Ashmore, Elizabeth Babister, and Ilan Kelman. "The Meaning of Build Back Better: Evidence from Post-Tsunami Aceh and Sri Lanka." *Journal of Contingencies and Crisis Management* 16.1 (2008): 24–36.

Kenny, Sue. "Reconstruction in Aceh: Building Whose Capacity?" *Community Development Journal* 42.2 (2007): 206–221.

Keohane, Robert O., and Joseph S. Nye. *Power and Interdependence*, 3rd ed. New York: Longman, (1977) 2001.

Kim, Jaeyop Y., Sookyung K. Park, and Clifton R. Emery. "The Incidence and

Impact of Family Violence on Mental Health Among South Korean Women: Results of a National Survey." *Journal of Family Violence* 24.3 (2009): 193–202.

King, C. Simrell, and Camilla Stivers. *Government Is Us: Public Administration in an Anti-Government era*. Thousand Oaks, CA: Sage Publications, 1998.

King, Dwight Y. "Civil service policies in Indonesia: An Obstacle to Decentralization?" *Public Administration and Development* 8 (1988): 249–260.

———. C. Simrell King, and Camilla Stivers. *Civil Service Policies in Indonesia: an Obstacle to Decentralization? Government Is Us: Public Administration in an Anti-Government Era*. Thousand Oaks, CA: Sage Publications, 1998.

Kirby, Peadar. "Theorising Globalisation's Social Impact: Proposing the Concept of Vulnerability." *Review of International Political Economy* 13.4 (2006): 632–655.

Koliba, Christopher, Russell Mills, and Asim Zia. "Accountability in Governance Networks: Implications Drawn from Studies of Response and Recovery Efforts Following Hurricane Katrina." *Public Administration Review* 71.2 (2011): 210–220.

Krause, Monika. "New Orleans: The Public Sphere of Disaster. In Understanding Katrina: Perspectives from the Social Sciences (Social Science Research Council)." June 11, 2006. http://understandingkatrina.ssrc.org/Krause/.

Krause, Sharon R. *Civil Passions: Moral Sentiment and Democratic Deliberation*. Princeton, NJ: Princeton University Press, 2008.

Kreps, Gary A. "Disasters and the Social Order." *Sociological Theory* 3 (1985): 49–65.

———, and Thomas E. Drabek. "Disasters as Nonroutine Social Problems." *International Journal of Mass Emergencies and Disasters* 14 (1996): 129–153.

Krieger, James, and Donna L. Higgin. "Housing and Health: Time Again for Public Health Action." *American Journal of Public Health* 92.5 (2002): 758–768.

Krugman, Paul. *The Great Unraveling: Losing Our Way in the New Century*. New York: Norton, 2003.

Kubicek, Paul. "The Earthquake, Civil Society, and Political Change in Turkey: Assessment and Comparison with Eastern Europe." *Political Studies* 50.4 (2002): 761–778.

Kukathas, Chandran. "Against the Communitarian Republic." *Australian Quarterly* 68 (1996): 67–76.

Kymlicka, Will. "Liberalism and Communitarianism." *Canadian Journal of Philosophy* 18 (1988): 181–203.

———. *Liberalism, Community and Culture*. Oxford, UK: Oxford University Press, 1989.

———. *Contemporary Political Philosophy: An Introduction*. Oxford, UK: Oxford University Press, 1990.

———. "Communitarianism, Liberalism, and Superliberalism." *Critical Review* 8 (1994): 263–284.

Landphair, Juliette. "The Forgotten People of New Orleans: Community, Vulnerability, and the Lower Ninth Ward." *Journal of American History* 94 (2007): 837–845.
Larmer, Brook. "Sichuan Earthquake, Poorly-Built Schools and Parents." *New York Times*, May 3, 2010.
Lawton Alan. *Ethical Management for the Public Services*. Buckingham, UK: Open University Press, 1998.
Le Grand, Julian. *The Other Invisible Hand: Delivering Public Services Through Choice and Competition*. Princeton, NJ: Princeton University Press, 2007.
Leon, Esteban, Ilan Kelman, James Kennedy, and Joseph Ashmore. "Capacity Building Lessons from a Decade of Transitional Settlement and Shelter." *International Journal of Strategic Property Management* 13.3 (2009): 247–265.
Lewis, Carol W. *The Ethics Challenge in Public Service*. San Francisco: Jossey-Bass Publishers, 1991.
Lewis, Judith A., Michael D. Lewis, Judy A. Daniels, and Michael J. D'Andrea. *Community Counseling: Empowerment Strategies for a Diverse Society*, 2nd ed. Pacific Grove, CA: Brooks/Cole, 1998.
———, M. Smith. Arnold R. House, and Rebecca L. Toporek. *ACA Advocacy Competencies*. 2002. http://www.counseling.org/Publications/. Last accessed March 2010.
Liu, Meiru. *Administrative Reform in China and Its impact on the Policy-Making Process and Economic Development After Mao: Reinventing Chinese Government*. Lewiston, NY: Edwin Mellen Press, 2001.
Longstaff, Patricia H., Nicholas J. Armstrong, Keli Perrin, Whitney M. Parker, and Matthew A. Hidek. "Building Resilient Communities: A Preliminary Framework for Assessment." *Homeland Security Affairs* 6.3 (2010): 1–23.
Lu, Yiyi. *Non-Governmental Organisations in China*. London: Routledge, 2009.
Lucia, Antoinette, and Richard Lepsinger. *The Art and Science of Competency Models: Pinpointing Critical Success Factors in Organizations*. San Francisco, CA: Jossey-Bass/Pfeiffer, 1999.
Luhmann, Niklas. *Risk: A Sociological Theory*, translated by Rhodes Barrett. New York: Aldine de Gruyter, 1993.
Luthar, Suniya S., and Dante Cicchetti. "The Construct of Resilience: Implications for Interventions and Social Policies." *Development and Psychopathology* 12 (2000): 857–885.
———, Dante Cicchetti, and Bronwyn Becker. "The Construct of Resilience: A Critical Evaluation and Guidelines for Future Work." *Child Development* 71 (2000): 543–562.
Maas, Arthur, and Laurence Radway. "Gauging Administrative Responsibility." *Public Administration Review* 19 (1959): 182–193.
MacIntyre, Alasdair. "How Moral Agents Became Ghosts or Why the History of Ethics Diverged from that of the Philosophy of Mind." *Synthese* 53 (1982): 295–312.
———. "Moral Philosophy: What Next?" In *Revisions: Changing Perspectives in*

Moral Philosophy, edited by Stanley Hauerwas and Alasdair MacIntyre, 1–15. Notre Dame, IN: University of Notre Dame Press, 1983.
———. "The Magic in the Pronoun 'My.'" *Ethics* 94 (1983): 113–125.
———. "Moral Rationality, Tradition, and Aristotle: A Reply to Onora O'Neill, Raymond Gaita, and Stephen R. L. Clark." *Inquiry* 26 (1983): 447–466.
———. *After Virtue*. London: Duckworth, 1984.
———. *After Virtue*, 2nd ed. Notre Dame, IN: University of Notre Dame Press, 1984.
———. *Whose Justice? Which Rationality?* Notre Dame, IN: University of Notre Dame Press, 1988.
———. "Plain Persons and Moral Philosophy: Rules, Virtues and Goods." *American Catholic Philosophical Quarterly* 66 (1992): 3–19.
———. "Critical Remarks on the Sources of the Self by Charles Taylor." *Philosophy and Phenomenological Research* 54 (1994): 187–190.
———. "A Partial Response to My Critics." In *After MacIntyre: Critical Perspectives on the Work of Alasdair MacIntyre*, edited by John Horton and Susan Mendus, 283–304. Notre Dame, IN: University of Notre Dame Press, 1994.
Makinen, Julie. "Japan Steps Up Nuclear Plant Precautions; Kan Apologizes." *L.A. Times*, March 25, 2011. http://www.latimes.com/news/nationworld/world/la-fgw-japan-nuclear-plant-20110326,0,5763742.story. Last accessed March 2012.
Mallik, Maggie. "Advocacy in Nursing—A Review of the Literature." *Journal of Advanced Nursing* 25.1 (1997): 130–138.
Mansfield, Richard S. "Building Competency Models: Approaches for HR Professionals." *Human Resource Management* 35 (1996): 7–18.
Manuel, John. "In Katrina's Wake." *Environmental Health Perspectives* 114.1 (2006): A32–39.
Marini, Frank, ed. "An Introduction: A New Public Administration?" In *Toward a New Public Administration: The Minnowbrook Perspective*, 17–48. San Francisco: Chandler Publishing Company, 1971.
Maxwell, Daniel, Sarah Bailey, Paul Harvey, Peter Walker, Cheyanne Sharbatki-Church, and Keven Savage. "Preventing Corruption in Humanitarian Assistance: Perceptions, Gaps and Challenges." *Disasters* 36.1 (2012): 140–160.
McAllister, Carol L., Tammy L. Thomas, Patrick C. Wilson, and Beth L. Green. "Root Shock Revisited: Perspectives of Early Heard Start Mothers on Community and Policy Environments and Their Effects on Child Health, Development and School Readiness." *American Journal of Public Health* 99.2 (2009): 205–210.
McCarthy James, Osvaldo. F. Canziani, Neil A. Leary, David J. Dokken, and Kasey C. White, eds. *Climate Change 2001: Impacts, Adaptation, and Vulnerability*: Contribution to Working Group II to the Third Assessment Report of the Intergovernmental Panel on Climate Change. Cambridge: Cambridge University Press, 2001.
McClelland, David C. "Testing for Competence Rather than for Intelligence." *American Psychologist* 28.1 (1973): 1–14.
McEntire, David A. "Sustainability or Invulnerable Development? Proposals

for the Current Shift in Paradigms." *Australian Journal of Emergency Management* 15.1 (2000): 58–61.

———. "Triggering Agents, Vulnerabilities and Disaster Reduction: Towards a Holistic Paradigm." *Disaster Prevention and Management* 10.3 (2001): 189–198.

———. Christopher Fuller, Chad W. Johnston, and Richard Weber. "A Comparison of Disaster Paradigms: The Search for a Holistic Policy Guide." *Public Administration Review* 62.3 (2002): 267–281.

McSwite, Orion C. *Legitimacy in Public Administration: A Discourse Analysis.* Thousand Oaks, CA: Sage Publications, 1997.

Menzel, Donald. "The Katrina Aftermath: A Failure of Federalism or Leadership?" *Public Administration Review* 66.6 (2006): 808–812.

Merriam-Webster's Collegiate Dictionary, 10th ed. Springfield, MA: Merriam-Webster, 1998.

Mileti, Dennis S. *Disasters by Design: A Reassessment of Natural Hazards in the United States.* Washington, DC: Joseph Henry Press, 1999.

———, Thomas E. Drabek, and J. Eugene Haas. *Human Systems in Extreme Environments: A Sociological Perspective.* Boulder, CO: Institute for Behavioral Science, University of Colorado, 1975.

Miller, David. *On Nationality.* Oxford, UK: Clarendon, 1995.

Miller, William R., and Stephen Rollnick. *Motivational Interviewing*, 2nd ed. London: Guilford Press, 2002.

Ministry of Civil Affairs of the People's Republic of China. "The General Review of Earthquake Disaster in China." July 14, 2008. http://cbzs.mca.gov.cn/article/sq/zbtj/200806/20080600016891.shtml. Last accessed March 2010.

Ministry of Personnel. Article 9, "Provisional Regulations on Civil Service Position Exchange." *China Personnel News*, August 31, 1996.

Mishra, Ramesh. *Globalization and the Welfare State.* Cheltenham, UK: Edward Elgar, 1999.

Moe, Terry M. "The Politicized Presidency." In *The New Direction in American Politics*, edited by John E. Chubb and Paul E. Peterson, 235–271. Washington, DC: Brookings Institution, 1985.

Mooney, Gerry. "'Problem' populations, 'problem' places." In *Social Justice: Welfare, Crime and Society,* edited by Janet Newman, and Nicole Yeates, 97–128. Maidenhead, UK: Open University Press, 2009.

Moore, Margaret. "Justice for Our Times." *Canadian Journal of Political Science* 23 (1990): 459–482.

———. *Foundations of Liberalism.* New York: Oxford University Press, 1993.

———. "Aceh Education Becomes Forgotten Casualty of War." *Sydney Morning Herald,* November 24, 2003. http://www.smh.com.au/articles/2003/11/23/1069522473591.html. Last accessed March 2010.

Mulhall, Stephen, and Adam Swift, *Liberals and Communitarians*, 2nd ed. Oxford, UK: Basil Blackwell, 1992.

Muramatsu, Michio. *Local Power in the Japanese State.* Berkeley: University of California Press, 1997.

———. "Postwar Politics in Japan: Bureaucracy Versus the Parties in Power." In *State and Administration in Japan and Germany*, edited by Michio

Muramatsu and Frieder Naschold, 13–38. New York: Walter de Gruyter, 1997.

Nagel, Thomas. *Equality and Partiality*. New York: Oxford University Press, 1991.

National Bureau of Statistics. *China Statistical Yearbook 2003*. Washington, DC: National Bureau of Statistics, 2003.

National Research Council (U.S.). *Committee on the Alaska Earthquake, The Great Alaska Earthquake of 1964*. Volume 1, Part 1. Washington, DC: National Academies, 1968.

Naudé, Wim, Mark McGillivray, and Stephanié Rossouw. "Measuring the Vulnerability of Subnational Regions in South Africa." *Oxford Development Studies* 37.3 (2009): 249–276.

Nazara, Suahasil, and Budy Resosudarmo. "Aceh-Nias Reconstruction and Rehabilitation: Progress and Challenges at the End of 2006." Tokyo: Asian Development Bank Institute, 2007. http://www.adbi.org/discussion-paper/2007/06/26/2288.acehnias.reconstruction.rehabilitation/. Last accessed June 2010.

Neal, David. "Reconsidering the Phases of Disaster." *International Journal of Mass Emergencies and Disasters* 15.2 (1997): 239–264.

Neal, Patrick, and David Paris. "Liberalism and the Communitarian Critique: A Guide for the Perplexed." *Canadian Journal of Political Science* 23 (1990): 419–439.

NESRI (National Economic and Social Rights Initiative). "International Advisory Group on Forced Evictions Investigates New Orleans Housing Crisis." Press release, July 21, 2009. http://www.nesri.org/AGFE_NOLA_July_2009_Press_Release.pdf. Last accessed March 2009.

———. *A Constant Threat: How HUD Policies and HANO Administration Deprive New Orleans' Residents of the Right to Housing*. New York: NESRI, 2010. http://www.nesri.org/sites/default/files/Community_Report_FINAL-Edited.pdf. Last accessed March 2009.

Nickel, Patricia M., and Angela M. Eikenberry. "Responding to 'Natural' Disasters: The Ethical Implications of the Voluntary State." *Administrative Theory & Praxis* 29.4 (2007): 534–545.

Nobles, Wade. *African Psychology: Toward Its Reclamation, Reascension & Revitalization*. Oakland, CA: A Black Family Institute Publication, 1986.

Norris, Fran H., Susan P. Stevens, Betty Pfefferbaum, Karen F. Wyche, and Rose L. Pfefferbaum. "Community Resilience as a Metaphor, Theory, Set of Capacities, and Strategy for Disaster Readiness." *American Journal of Community Psychology* 41.1–2 (2008): 127–150.

Nussbaum, Martha. "The Discernment of Perception: An Aristotelian Conception of Private and Public Rationality." In *Love's Knowledge: Essays on Philosophy and Literature*, 54–105. New York, Oxford: Oxford University Press, 1990.

Nussbaum, Martha C. *Women and Human Development: The Capabilities Approach*. Cambridge: Cambridge University Press, 2000.

———. "Capabilities as Fundamental Entitlements: Sen and Social Justice." *Feminist Economics* 9.2–3 (2003): 33–59.

Nye, Joseph S. "Corruption and Political Development: A Cost-Benefit Analysis." *American Political Science Review* 61 (1967): 417–427.

O'Brien, Robert, and Marc Williams. *Global Political Economy: Evolution and Dynamics*. Basingstoke: Palgrave MacMillan, 2004.
Office of the State Council. "Notice on Enhancing Management and Use of Wenchuan Earthquake Disaster Relief Funds and Materials." http://www.gov.cn/xxgk/pub/govpublic/mrlm/200806/t20080602_32849.html.
Okin, Susan M. *Justice, Gender, and the Family*. New York: Basic Books, 1989.
Oldenquist, Andrew. "Loyalties." *Journal of Philosophy* 79.4 (1982): 173–93.
O'Leary, Rosemary. *The Ethics of Dissent. Managing Guerrilla Government*. Washington, DC: CQ Press, 2006.
Oxfam. "Hurricane Dennis Leaves Behind Destruction in Cuba." *Oxfam News*, July 14, 2005. http://www.oxfam.ca/news/hurricane_dennis/july14update.htm.
Özerdem, Alpaslan, and Tim Jacoby. *Disaster Management and Civil Society: Earthquake Relief in Japan, Turkey and India*. London: I. B. Tauris, 2006.
Paton, Douglas. *Measuring and Monitoring Resilience in Auckland*. GNS Science Report, 2007/18. New Zealand: GNS Science, 2007.
———, Leigh Smith, and John Violanti. "Disaster Response: Risk, Vulnerability and Resilience." *Disaster Prevention and Management* 9.3 (2000): 173–179.
Paul, Jeffrey, and Fred D. Miller, Jr. "Communitarian and Liberal Theories of the Good." *Review of Metaphysics* 43 (1990): 803–830.
Pekkanen, Robert. "Japan's New Politics: The Case of the NPO Law." *Journal of Japanese Studies* 26.1 (2000): 111–148.
Pelling, Marc, and Chris High. "Understanding Adaptation: What Can Social Capital Offer Assessment of Adaptive Capacity?" *Global Environmental Change* 15.4 (2005): 308–319.
Petal, Marla, Rebekah Green, Ilan Kelman, Rajib Shaw, and Amod Dixit. "Community-Based Construction for Disaster Risk Reduction." In *Hazards and the Built Environment*, edited by Lee L. Bosher, 191–217. London: Taylor and Francis, 2008.
Peter, Fabienne. "Political Equality of What? Deliberative Democracy, Legitimacy and Equality." Paper for the 2004 Annual Meeting of the American Political Science Association, 2–5 September 2004, http://www.apsanet.org.
Pietz, David. *Engineering the State: The Huai River and Reconstruction in Nationalist China 1927–1937*. London: Routledge, 2002.
Pogge, Thomas W. "Can the Capability Approach Be Justified?" *Philosophical Topics* 30.2 (2002): 167–228.
Pokorny, Jaroslav, and John Storek. "Current Development in the Czech Republic's EMS in the Aftermath of Big Catastrophes." *Acta* 44 (2002):115–116.
Polanyi, Karl. "Our Obsolete Market Mentality." In *Primitive, Archaic and Modern Economies: Essays of Karl Polanyi*, edited by George Dalton, 59–77. New York: Anchor Books, 1968.
———. *The Livelihood of Man*. New York: Academic Press, 1977.
———. *The Great Transformation*. Boston: Beacon Books, (1944) 2001.
Pops, Gerald. "A Teleological Approach to Administrative Ethics." In *The Handbook of Administrative Ethics*, edited by Terry Cooper, 157–168. New York: M. Decker Publisher, 1994.

Potter Nancy. *How Can I Be Trusted? A Virtue Theory of Trustworthiness.* Lanham, MD: Rowan and Littlefield, 2002.

"Protesters Attack Aceh Tsunami Reconstruction Office." *Reuters*, September 21, 2006.

Pugh, Darrell L. "The Origins of Ethical Frameworks in Public Administration." In *Ethical Frontiers in Public Management*, edited by James S. Bowman, 9–33. San Francisco: Jossey-Bass Publishers, 1991.

Putnam, Robert. "The Prosperous Community: Social Capital and Public Life." *The American Prospect* 13 (1993): 35–41.

———. "Bowling Alone: America's Declining Social Capital." *Journal of Democracy* 6.1 (1995): 65–78.

———. *Bowling Alone: The Collapse and Revival of American Community.* New York: Simon and Schuster, 2000.

Quarantelli, Enrico L. "Statistical and Conceptual Problems in the Study of Disasters." *Disaster Prevention and Management*, 10.5 (2001): 325–338.

———. *What Is Disaster? Perspectives on the Question.* London: Routledge, 1998.

———. "Urban Vulnerability and Technological Hazards in Developing Societies." In *Environmental Management and Urban Vulnerability*, edited by Alcira Kreimer and Mohan Munasinghe, 187–236. Washington, DC: World Bank, 1992.

Quill, Lawrence. "Ethical Conduct and Public Service: Loyalty Intelligently Bestowed." *The American Review of Public Administration* 39 (2009): 215–224.

Radice, Betty, trans. *The Letters of the Younger Pliny.* London: Penguin Books, 1969.

Rappaport, Julian. "In Praise of Paradox: A Social Policy of Empowerment over Prevention." *American Journal of Community Psychology* 9 (1981): 1–25.

Rasheed, Aesha. "Chaos and Hope in New Orleans Schools, One Year After Katrina." *Southern Exposure* special report, XXXIV.2 (2006): 1–100.

Rawls, John. *A Theory of Justice.* Cambridge, MA: Harvard University Press: 1971.

Redding, Gordon S. *The Spirit of Chinese Capitalism.* Berlin: de Gruyter, 1990.

Reich, Robert. *The Power of Public Ideas.* Cambridge: Ballinger Publishing Company, 1988.

Riely, Frank. "A Comparison of Vulnerability Analysis: Methods and Rationale for Their Use in Different Contexts." 2000. http://www.fivims.net/documents/RielyVGProfilingMethodsAnnex.doc. Last accessed March 2009.

Ritea, Steve. "Public schools' makeup similar." 2006a. http://www.nola.com/education. Last accessed May 2010.

———. "Skeleton Crew Left to Gut N.O. System." January 21, 2006b. http://www.nola.com/education. Last accessed May 2010.

———. "School district pledges to end wait lists." 2007. http://www.nola.com/education. Last accessed May 2010.

Roberts, Norman J., Farokkh Nadim, and Bjørn Kalsnes. "Quantification of Vulnerability to Natural Hazards." *Georisk* 3.3 (2009): 164–173.

Robeyns, Ingrid. "Sen's Capability Approach and Gender Inequality: Selecting Relevant Capabilities." *Feminist Economics* 9.2–3 (2003): 61–92.

Rodan, Garry, and Caroline Hughes. "Ideological Coalitions and the International Promotion of Social Accountability: The Philippines and Cambodia Compared." *International Studies Quarterly* 56 (2012): 367–380.

Rodriguez, Donna, Rita Patel, Andrea Bright, Donna Gregory, and Marilyn K. Gowing. "Developing Competency Models to Promote Integrated Human Resource Practices." *Human Resource Management* 41 (2002): 309–24.

Romzek, Barbara S., and Melvin J. Dubnick. "Accountability in the Public Sector: Lessons from the Challenger Tragedy." In *Democracy, Bureaucracy, and Administration*, edited by Camilla Stivers, 182–204. Boulder, CO: Westview Press, 2001.

Room, Graham. *Beyond the Threshold: The Measurement and Analysis of Social Exclusion*. Bristol, UK: The Policy Press, 1995.

Rosenstein, Carole. "New Orleans Arts and Culture." In *After Katrina: Shared Challenges for Rebuilding Communities*, edited by Carol J. DeVita, 13–16. Washington, DC: The Urban Institute; October 20, 2007. http://www.urban.org/UploadedPDF/311440_After_Katrina.pdf. Last accessed May 2010.

Ross, Michael L. "Resources and Rebellion in Aceh, Indonesia." In *Understanding Civil War: Evidence and Analysis*, edited by Paul Collier and Nicholas Sambanis, 35–58. Washington: The World Bank, 2005.

Rourke, Francis E. Responsiveness and Neutral Competence in American Bureaucracy. *Public Administration Review* 52(6) (1992): 539-546.

Ruscher, Janet B. "Stranded by Katrina: Past and Present." *Analyses of Social Issues and Public Policy* 6.1 (2006): 33–38.

Rutter, Michael. "Resilience Concepts and Findings Implications for Family Therapy." *Journal of Family Therapy* 21 (1999): 119–144.

Sabatier, Paul. "Top-Down and Bottom-Up Models of Policy Implementation: A Critical Analysis and Suggested Synthesis." *Journal of Public Policy* 6 (1986): 21–48.

Saith, Ruhi. "Capabilities: The concept and Its Operationalisation." Queen Elizabeth House Working Paper Series no. 66, University of Oxford, 2001.

Sandel, Michael. *Liberalism and the Limits of Justice*. Cambridge: Cambridge University Press, 1982.

———. "The Procedural Republic and the Unencumbered Self." *Political Theory* 12.1 (1984): 81–96.

———. "Moral Argument and Liberal Toleration: Abortion and Homosexuality." In *New Communitarian Thinking: Persons, Virtues, Institutions, and Communities*, edited by Amitai Etzioni, 71–87. Charlottesville, VA: University Press of Virginia, (1989) 1995.

———. "Political Liberalism." *The Harvard Law Review* 107 (1994): 1765–1794.

———. *Democracy's Discontent: America in Search of a Political Philosophy*. Cambridge: Harvard University Press, 1996.

Santacroce, Roberto. "A General Model for the Behavior of the Somma

Vesuvius Volcanic Complex." *Journal of Volcanology and Geothermal Research* 17 (1983): 237–248.
Scheffler, Samuel. *The Rejection of Consequentialism.* New York: Oxford University Press, 1982.
———. *Boundaries and Allegiances.* Oxford, UK: Oxford University Press, 2001.
Schmitter, Philippe C. "Civil Society and Democratization." In *INPR, International Conference on Consolidating the Third Wave Democracies: Trends and Challenges*, 59–69. Taipei: Institute for National Policy Research, 1995.
Scholte, Jan Aart. *Globalization: A Critical Introduction*, Basingstoke: Palgrave, 2000.
Schoon Ingrid. *Risk and Resilience: Adaptations in Changing Times.* Cambridge, UK: Cambridge University Press, 2006.
Schott, Christina. "'Smong' Legend Becomes a Lifesaver: Sumatra, Indonesia." *Earthquake Spectra* 22.S3, April 4, 2006.
Schwartz, Jonathan, and Shawn Shieh, eds. *State and Society Responses to Social Welfare Needs in China: Serving the People.* London: Routledge, 2009.
Selznick, Philip. "The Idea of a Communitarian Morality." *California Law Review* 75 (1987): 445–463.
Sen, Amartya. *Poverty and Famines: An Essay on Entitlement and Deprivation.* Oxford, UK: Clarendon Press, 1981.
Sen, Amartya K. *Commodities and Capabilities.* Amsterdam: North Holland, 1985.
———. "Capability and Well-Being." In *The Quality of Life*, edited by Martha Nussbaum and Amartya Sen, 30–53. Oxford: Clarendon Press, 1993.
———. *Development as Freedom.* New York: Anchor Books, 1999.
Sheridan, Michael F., Franco. Barberi, Mauro Rosi, and Roberto Santacroce. "A Model of Plinian Eruptions of Vesuvius." *Nature* 289 (1981): 282–228.
Shieh, Shawn, and Jonathan Schwartz. "State and Society Responses to Social Welfare Needs in China: An Introduction to the Debate." In *State and Society Responses*, edited by Jonathan Schwartz and Shawn Shieh. Abingdon, UK: Routledge, 2010.
Shreib, Kathleen, Frank Norris, and Sandro Galea. "Measuring Capacities for Community Resilience." *Social Indicators Research* 99.2 (2010): 227–247.
Sieg, Linda. "Japan Government Losing Public Trust as Nuclear Crisis Worsens." *Reuters* March 15, 2011. http://www.chinapost.com.tw/asia/japan/2011/03/16/294894/p1/Japan-government.html. Last accessed May 2010.
Sigurdsson, Haraldur, Steven Carey, Winton Cornell, and Tullio Pescatore. "The Eruption of Vesuvius in A.D. 79." *National Geographic Research* 1.3 (1985): 332–387.
Silver, Hilary. "Reconceptualizing Social Disadvantage: Three Paradigms of Social Exclusion." In *Social Exclusion: Rhetoric, Reality, Responses*, edited by Gerry Rodgers, Charles Gore, and Jose B. Figueiredo, 57–80. Geneva: International Institute for Labour Studies, 1995.

Simon, Darran. "New School Era Opens Today." September 4, 2007. http://www.nola.com/education.

Skowronek, Stephen. *Building a New American State: The Expansion of National Administrative Capacities 1877–1920.* Cambridge, UK: Cambridge University Press, 1982.

Smith, Steven Rathgeb. "The Challenge of Strengthening Nonprofits and Civil Society." *Public Administration Review* 68 (2008): S132–145.

Solnit, Rebecca. *A Paradise Built in Hell: The Extraordinary Communities that Arise in Disaster.* New York: Viking, 2009.

Souza, Renato, Sasha Bernatsky, Rosalie Reyes, and Kaz de Jong. "Mental Health Status of Vulnerable Tsunami-Affected Communities: A Survey in Aceh Province, Indonesia." *Journal of Traumatic Stress* 20.3 (2007): 263–269.

Sovjakova, Eva. "Floods in July 1997 (Czech Republic)." In *NEDIES Project Lessons Learnt from Flood Disasters*, edited by Alessandro Colombo and Ana Lisa Vetere Arellano. Report EUR 20261 EN. Italy: European Commission, 2002.

Spake, Amanda. "Dying for a Home." *The Nation* February 14, 2007. http://www.thenation.com/doc/20070226/spake. Last accessed May 2011.

Spence, Jonathan. *The Search for Modern China.* New York: W. W. Norton & Company, 1991.

Spencer, Lyle M., and Signe G. Spencer. *Competence at Work: Models for Superior Performance.* New York: Wiley, 1993.

Spicer, Michael W., and Larry D. Terry. "Legitimacy, History, and Logic: Public Administration and the Constitution." *Public Administration Review* 53 (1993): 239–246.

State Council State Council Information Office via SINA.com. 2008-09-04. http://news.sina.com.cn/c/2008-09-04/105416231615.shtml. Last accessed May 2010.

Stepan, Alfred. *Rethinking Military Politics.* Princeton, NJ: Princeton University Press, 1988.

Stephenson, Max, Jr., and Marcey H. Schnitzer. "Interorganizational Trust, Boundary Spanning, and Humanitarian Relief Coordination." *Nonprofit Management and Leadership* 17.2 (2006): 211–233.

Stewart, Frances, and Severine Deneulin. "Amartya Sen's Contribution to Development Thinking." *Studies in Comparative International Development* 37.2 (2002): 61–70.

Stiglitz, Joseph. *The Roaring Nineties: Seeds of Destruction,* London: Allen Lane, 2003.

Stivers, Camilla. "The Public Agency as Polis: Active Citizenship in the Administrative State." *Administration and Society* 22 (1990): 86–106.

———. "The Listening Bureaucrat: Responsiveness in Public Administration." *Public Administration Review* 54 (1994): 364–369.

———. "So Poor and so Black: Hurricane Katrina, Public Administration, and the Issue of Race." *Public Administration Review* Special Issue, December (2007): 48–56.

———. *Governance in Dark Times: Practical Philosophy for Public Service.* Washington, DC: Bristol: Georgetown University Press, 2008.

Streich, Gregory W. "Constructing Multiracial Democracy: To Deliberate or Not to Deliberate?" *Constellations* 9.1 (2002): 127–153.
Sudmeier-Rieux, Karen, Hillary Masundire, Ali Rizvi, and Simon Rietbergen. *Ecosystems, Livelihoods and Disasters: An Integrated Approach to Disaster Risk Management*. Gland and Cambridge, UK: IUCN, 2006.
Sulaiman, M. Isa. "From Autonomy to Periphery: A Critical Evaluation of the Acehnese Nationalist Movement." In *Verandah of Violence*, edited by Anthony Reid. Singapore: National University of Singapore, 2006.
Sutinen, Jon G., and Karen Kuperan. "A Socio-Economic Theory of Regulatory Compliance." *International Journal of Social Economics* 26.1–3 (1999): 174–193.
Svara, James. *The Ethics Primer for Public Administrators in Government and Nonprofit Organizations*. Sudbury, MA: Jones & Bartlett, 2007.
Syarif, Sanusi M. *Gampong dan Mukim di Aceh: Menuju Rekonstruksi Paska Tsunami*. Bogor: Pustaka Latin, 2005.
Tacitus. *Annals*, XV.39. Edited by Alfred John Church, William Jackson Brodribb, and Sara Bryant. New York: Random House, 1942.
———. *Annals*, XV 40. Edited by Alfred John Church, William Jackson Brodribb, and Sara Bryant. New York: Random House, 1942.
Tamir, Yael. *Liberal Nationalism*. Princeton, NJ: Princeton University Press, 1993.
Tayag, Jean, Sheila Insauriga, Anne Ringor, and Mel Belo, "People's Response to Eruption Warning: The Pinatubo Experience, 1991–92." In *Fire and Mud: Eruptions and Lahars of Mount Pinatubo, Philippines, Quezon City*, edited by Chris G. Newhall and Ray S. Punongbayan. Seattle and London: Philippine Institute of Volcanology and Seismology and University of Washington Press, 1996.
Taylor, Charles. "Hegel: History and Politics." In *Liberalism and Its Critics*, edited by Michael Sandel, 177–199. New York: New York University Press, 1984.
———. *Philosophical Papers*. Cambridge, UK: Cambridge University Press, 1985.
———. "Alternative Futures: Legitimacy, Identity and Alienation in Late Twentieth Century Canada." In *Constitutionalism, Citizenship and Society in Canada*, edited by Alan Cairns and Cynthia Williams, 183–229. University of Toronto Press, Toronto, 1985.
———. *Sources of the self: The making of moral identity*. Cambridge, UK: Cambridge University Press, 1989.
———. "Justice After Virtue." In *After MacIntyre*, edited by John Horton and Susan Mendus, 16–44. Cambridge: Polity Press, 1994.
———. "Cross-Purposes: The Liberal-Communitarian Debate." In *Philosophical Arguments*. Cambridge, MA: Harvard University Press, 1995.
Teets, Jessica. "Post-Earthquake Relief and Reconstruction Efforts: The Emergence of Civil Society in China?" *The China Quarterly* 198 (2009): 330–347.
Telford, John. "Learning Lessons from Disaster Recovery: The Case of Honduras." World Bank Disaster Risk Management Working Paper Series No. 8, 2004

Terry, Larry D. "The Thinning of Administrative Institutions in the Hollow State." *Administration & Society* 37 (2005): 426–444.
Thomas, Alan. "Reasonable Partiality and the Agent's Point of View." *Ethical Theory and Moral Practice*, 8 (2005): 25–43.
Thompson, Martha, and Izaskun Gaviria. *Weathering the Storm: Lessons in Risk Reduction from Cuba*. Boston, MA: Oxfam America, 2004.
Thorburn, Craig. "The Acehnese Gampong Three Years On: Assessing Local Capacity & Reconstruction in Post-Tsunami Aceh." Report of the Aceh Community Assistance Research Project (ACARP), 2007. http://www.indo.ausaid.gov.au/featurestories/acarpreport.pdf. Last accessed May 2011.
Thornton, Patricia M. "Crisis and Governance: SARS and the Resilience of the Chinese Body Politic." *The China Journal* 61 (2009): 23–48.
Tomasello, Michael. *The Cultural Origins of Human Cognition*. Cambridge, MA: Harvard University Press, 1999.
Trim, Peter R. J. "An Integrative Approach to Disaster Management and Planning." *Disaster Prevention and Management* 13.3 (2004): 218–222.
Tsunami Evaluation Coalition. *Impact of the Tsunami Response on Local and National Capacities. Indonesia Country Report (Aceh and Nias)*. Elizabeth Scheper with contributions from Smruti Patel and Arjuna Parakrama, London: Tsunami Evaluation Committee (TEC), April 2006.
Tuzzolo, Ellen, and Damon T. Hewitt. "Rebuilding Inequity: The Re-emergence of the School-to-Prison Pipeline in New Orleans." *The High School Journal* 90.2 (2006): 59–68
Tyack, David, and Larry Cuban. *Tinkering Towards Utopia: A Century of Public School Reform*. Cambridge, MA: Harvard University Press, 1995.
UDHR. Universal Declaration of Human Rights. G.A. Res. 217A (III) (1948), Art. 25, Sec. 1. http://www.un.org/Overview/rights.html.
UN. Article 5, Section 3 of the International Convention on the Elimination of All Forms of Racial Discrimination. http://www2.ohchr.org/english/law/cerd.htm. Last accessed May 2011.
———. Article 9, American Declaration of the Rights and Duties of Man, O.A.S. Res. XXX, adopted by the Ninth International Conference of American States (1948). http://www1.umn.edu/humanrts/oasinstr/zoas2dec.htm. Last accessed May 2011.
———. *Report on the World Social Situation: Social Vulnerability: Sources and Challenges*. New York: United Nations Department of Economic and Social Affairs, 2003.
UNDP (United Nations Development Programme). *Human Development Report 1999*. New York: Oxford University Press, 1999.
———. *Evolution of a Disaster Risk Management System: A Case study from Mozambique*. Geneva, Switzerland: United Nations Bureau for Crisis Prevention and Recovery, 2004.
———. *Civil Society in Aceh: An Assessment of Needs to Build Capacity to Support Community Recovery*. New York: UN, 2005.
UNHCR (United Nations High Commissioner for Refugees). *Handbook for Emergencies*. Geneva, Switzerland, 2007.

UNEP. "Global Environment Outlook 3." 2003. www.unep.org/GEO/geo3/. Last accessed May 2011.
UNISDR. *Living with Risk: A Global Review of Disaster Reduction Initiatives.* United Nations, 2004.
U.S. Department of Homeland Security. *National Preparedness Guidelines.* Washington, DC, 2007.
U.S. Department of Labor, Bureau of Labor Statistics. "Federal Government, Excluding the Postal Service." Section: Employment. March 12, 2008. http://www.bls.gov/oco/cg/cgs041.htm. Last accessed May 2011.
U.S. Department of the Interior. "The Federal Civil Service." (5 U.S.C. § 2101). DOI University, National Business Center, Revised October 11, 1998. http://www.doi.gov/hrm/pmanager/st6.html. Last accessed May 2011.
Van Wart, Montgomery. *Changing Public Sector Values.* New York: Garland Publishing, 1998.
Varley, Ann. "The Exceptional and the Everyday: Vulnerability Analysis in the International Decade for Natural Disaster Reduction." In *Disasters, development, and environment,* edited by Ann Varley, 1–11. New York: John Wiley & Sons, 1994.
Vazirani, Nitin. "Competencies and Competency Model—A Brief Overview of Its Development and Application." *SIES Journal of Management* 7.1 (2010): 121–131.
Ventriss, Curtis. "Two Critical Issues of American Public Administration: Reflections of a Sympathetic Participant." *Administration & Society* 19 (1987): 25–47.
Vung, Naguyen D., Per-Olof Ostergren, and Gunilla Krantz. "Intimate Partner Violence Against Women, Health Effects and Health Care Seeking in Rural Vietnam." *European Journal of Public Health* 19.2 (2009): 178–182.
Wade, Robert Hunter. "The Disturbing Rise in Poverty and Inequality: Is It All a 'Big Lie'?" In *Taming Globalization: Frontiers of Governance,* edited by David Held and Mathias Koenig-Archibugi, 18–46. Cambridge, UK: Polity Press, 2003.
Waldo, Dwight. *The Enterprise of Public Administration: A Summary View.* Novato, CA: Chandler and Sharp Publishers, 1981.
Walker, Alan C., ed. *Britain Divided: The Growth of Social Exclusion in the 1980s and 1990s.* London: Child Poverty Action Group, 1997.
Wallace, Deborah, and Rodrick Wallace. *A Plague on Your Houses.* London/New York: Verso, 1998.
Walsh, Froma. "A Family Resilience Framework: Innovative Practice Applications." *Family Relations* 51.2 (2002): 130–137.
Walzer, Michael. *Spheres of justice.* New York: Basic Books, 1983.
———. *Interpretation and Social Criticism.* Cambridge: Harvard University Press; 1987.
———. "Socializing the Welfare State." *Dissent* Summer (1988): 292–300.
———. "The Idea of Civil Society." *Dissent* Spring (1991): 293–304.
———. "Response." In *Pluralism, Justice, and Equality,* edited by David Leslie Miller and Michael Walzer, 281–298. Oxford: Oxford University Press, 1995.

———. "Michael Sandel's America." In *Debating Democracy's Discontent: Essays on American Politics*, edited by Anita L. Allen and Milton C. Regan, Jr., 175–182. Oxford: Oxford University Press, 1998.
———. "Seminar with Michael Walzer." *Ethical Perspectives* 6.3–4 (1999): 220–242.
———. "Equality and Civil Society." In *Alternative Conceptions of Civil Society*, edited by Simone Chambers and Will Kymlicka, 34–49. Princeton, NJ: Princeton University Press, 2002.
Wang, Li, Yuqing Q. Zhang, Wenzhong Z. Wang, Zhanbiao B. Shi, Jianhua H. Shen, and Ming Li. "Symptoms of Posttraumatic Stress Disorder Among Adult Survivors Three Months After the Sichuan Earthquake in China." *Journal of Traumatic Stress* 22.5 (2009): 444–450.
Wang Ming, See. "The Development and Present Situation of NGOs in China." *Social Sciences in China* 28.2 (2007): 99–100.
———. Tao Chuanjing, and Han Junkui, eds. *Wenchuan dizhen gongmin xingdong baogao* (Report on Civil Society Action in the Wenchuan Earthquake). Beijing: Social Science Academic Press, 2008.
Warner, Charles. "Demographer Says Many Residents Want to Return." February 12, 2005. http://www.nola.com. Last accessed May 2011.
Washington, Henry S. "Postscript." In *The Vesuvius Eruption of 1906: Study of a Volcanic Cycle*, edited by Frank Alvord Perret, 97–98. Publ. 339. Washington, DC: Carnegie Institute, 1924.
Watson, Andrew. "Civil Society in a Transitional State: The Rise of Associations in China." In *Associations and the Chinese State: Contested Spaces*, edited by Jonathan Unger. Armonk, NY: M. E. Sharpe, 2008.
Watson, Robert T., Marufu C. Zinyowera, and Richard H. Moss, eds. *Climate Change 1995—Impacts, Adaptations and Mitigation of Climate Change: Scientific-Technical Analyses*. Cambridge, UK: Cambridge University Press, 1996.
Watts, Michael J., and Hans-Georg Bohle. "The Space of Vulnerability: The Causal Structure of Hunger and Famine." *Progress in Human Geography* 17.1 (1993): 43–67.
Waugh, William L., Jr. *Living with Hazards, Dealing with Disaster: An Introduction to Emergency Management*. Armonk, NY: M. E. Sharpe., 2000.
———. "EMAC, Katrina, and the Governors of Louisiana and Mississippi." Special issue, *Public Administration Review* 67 (2007): 107–113.
Weaver, Matthew. "Police Break Up Protest by Parents of China Earthquake Victims." *Guardian*, June 3, 2008.
Webb, Patrick, and Anuradha Harinarayan. "A Measure of Uncertainty: The Nature of Vulnerability and Its Relationship to Malnutrition." *Disasters* 23.4 (1999): 292–305.
Weil, Frederick. "The Rise of Community Engagement After Katrina." In *The New Orleans Index at Five*. Washington, DC: Brookings Institution and Greater New Orleans Community Data Center, 2010.
White, Gilbert F. *Human Adjustment to Floods: A Geographical Approach to the Flood Problem in the United States*. Chicago, IL: Department of Geography, University of Chicago, 1945.

———. *Choice of Adjustment to Floods*. Chicago, IL: University of Illinois Press, 1964.

———, ed. *Natural Hazards: Local, National, Global*. New York: Oxford University Press, 1974.

White, Gordon, Jude Howell, and Xiaoyuan Shang. *In Search of Civil Society: Market Reform and Social Change in Contemporary China*. Oxford, UK: Clarendon Press, 1996.

WHO Commission on Social Determinants of Health. *Closing the Gap in a Generation: Health Equity Through Action on Social Determinants of Health*. Geneva: WHO, 2008.

———. *Knowledge Network on Urban Settings, Our Cities, Our Health, Our Future: Acting on Social Determinants for Health Equity in Urban Settings: Final Report of the Urban Settings Knowledge Network*. Geneva: WHO, 2008.

Wilder, Andrew. "Aid and Stability in Pakistan: Lessons from the 2005 Earthquake Response." *Disasters* 34(s3) (2010): S406–426.

Williams, Bernard A. O. "Persons, Character, and Morality." In *Identities of Person*, edited by Amélie Oksenberg Rorty, 197–217. Berkeley, CA: University of California Press, 1974. Reprinted in Williams, B. A. O., *Moral Luck*. Cambridge, UK: Cambridge University Press, 1981.

Williams, Martyn. "Japan Expands Evacuation Zone Around Nuclear Plant." VOANews.com. April 22, 2011. http://www.voanews.com/english/news/asia/east-pacific/Japan-Expands-Evacuation-Zone-Around-Fukushima-Daiichi-Plant-120445924.html.

Winchester, Simon. *The River at the Center of the World: A Journey Up the Yangtze, and Back in Chinese Time*. Oxford, UK: Macmillan, 2004. Last accessed May 2011.

Wolf, Ernest S. *Treating the Self: Elements of Clinical Self-Psychology*. New York: Guilford Press, 1988.

Woller, Gary M., and Kelly D. Patterson. "Public Administration Ethics: A Postmodern Perspective." *American Behavioral Scientist* 41 (1997): 103–118.

Wong, Edward. "China Presses Hush Money on Grieving Parents." *New York Times*, July 24, 2008. http://www.nytimes.com/2008/07/24/world/asia/24quake.html. Last accessed May 2011.

"Words into Action: A Guide for Implementing the Hyogo Framework." www.unisdr.org/eng/hfa/docs/Words-into-action/Words-Into-Action.pdf, p. 5. Last accessed May 2011.

World Bank. *World Development Report 2000-01: Attacking Poverty*. New York: Oxford University Press, 2000.

———. "Pay and Patronage in the Core Civil Service in Indonesia." Mimeo, World Bank. http://www1.worldbank.org/publicsector/civilservice/countries/indonesia/shapesize.htm. Last accessed May 2011.

———. *Globalization, Growth, and Poverty: Building an Inclusive World economy*. New York: Oxford University Press with the World Bank, 2002.

World Commission on Environment and Development (Brundtland Commission). *Our Common Future*. Oxford, UK: Oxford University Press, 1987.

Xinhua News Agency. "Beijing, Jiangxi, Hunan, Inner Mongolia, Anhui,

Shanghai, Li Ning, Hei Nong Jiang, and Shengzhen Organize Post Quake Emergency Groups and Relief Teams to Response to Resources and Rehabilitation Needs." May 21, 2008. http://big5.xinhuanet. com/gate/big5/news.xinhuanet.com/newscenter/2008-05/21/ content_8222881.htm. Last accessed May 2011.

Yang, Guobin. "A Civil Society Emerges from the Earthquake Rubble." Yale Global Online, www.yaleglobal.com, June 5, 2008.

Young, Marion Iris. *Inclusion and Democracy.* Oxford, UK: Oxford University Press: 2000.

Young, Nick. "250 Chinese NGOs: Civil Society in the Making." Beijing: China Development Brief, 2001.

Yuanchang, Zheng. "Lessons Modern Chinese Charity Workers Can Take from Society's Response to the Wenchuan Earthquake." *China Nonprofit Review* 3 (2008): 130–141.

Zanetti, Lisa A. "Advancing Praxis: Connecting Critical Theory with Practice." *The American Review of Public Administration* 27 (1997): 145–167.

Zhuang, Tianhui, Haixia Zhang, Guopei Zhang, and Xiaoping Zheng. "Research on the Influencing Factors of Peasants' Autonomy to Play in Post-Earthquake Reconstructions—A Case Study of Wenchuan Earthquake in China." *Journal of Sustainable Development* 3.3 (2010): 175–183.

Index

Page numbers in *italics* indicate figures; those with a *t* indicate tables.

Aceh tsunami (2004), 8–9, 104, 124–36, 171–73
Action against Child Exploitation (ACE), 158
Advisory Group on Forced Evictions (AGFE), 111
advocacy in community-based resilience management, 81, *82,* 94–97; Aceh tsunami and, 124–27, 171–72; Fukushima Daiichi nuclear accident and, 153–56, 172; guidelines for, 174–76; Hurricane Katrina and, 106–12, 171, 172; professional helpers and, 166–67, 171; Wenchuan earthquake and, 138–42, 172
Advocacy Services for Justice and Reconciliation (Indonesia), 135
Aguirre, Benigno, 50
Ai Weiwei, 140
Alaskan earthquake and tsunami (1964), 20
Aldrich, Daniel, 132
American Nurses Association (ANA), 95
Anastario, Anastario, 142
Ansombe, Elizabeth, 67
Aristotle, 67, 69
arts and cultural communities: Aceh tsunami and, 131–32; Fukushima Daiichi nuclear accident and, 162–63; Hurricane Katrina and, 117–19; Wenchuan earthquake and, 150, 172
Asian Friendship Society, 157
Association for Aid and Relief (Japan), 162
Association of Medical Doctors of Asia, 161

Badan Keswadayaan Masyarakat, 134
Bangladesh cyclone (1970), 21
Barrows, Harlan H., 31
Beatley, Timothy, 52
Beijing Pengbo Cultural Training School, 150
Beiner, Ronald, 74
Bellah, Robert N., 73
Berke, Phillip R., 54
Bhalla, Ajit, 100
Bhola cyclone (1970), 21
Blaikie, Piers, 36
Bohle, Hans-Georg, 35
Boudreau, Tania, 37
Boyatzis, Richard, 97–98
Brazil, 41, 52
Bridge Asia Japan, 160
Broadmoor Improvement Association, 110
Brooks Education Center (Beijing), 149
Brundtland Commission Report, 51–52

Buckle, Philip, 55
Burke, Edmund, 69
Burton, Ian, 31
Bush, George W., 107, 115–17

Campaign to Restore National Housing Rights, 111
CARE International, 162
Caring for Young Refugees, 158–59
Center for Public Integrity, 113
Challenger Space Shuttle Disaster (1986), 77–78
Chambers, Robert, 33–35
Chandler, Ralph, 65
charter schools, 115–16
Chengdu Education Foundation, 147–48
Chernobyl nuclear accident, 156
Chicago-Colorado-Clark-Toronto School of Natural Hazard Studies, 31
Child Fund (organization), 157–58
Child Survival Kits, 129
Children without Borders (Japan), 162
children's issues, 127–29, 148, 156–62. *See also* women's issues
Chilean earthquake (1647), 13
China: civil service of, 123–24, 136–37; Tangshan earthquake in, 21–22; Wenchuan earthquake in, 8–9, 104, 138–51, 171–73; Yangtze River flood in, 20
China Charity Federation, 146
China Children and Teenager's Fund, 148
China Education Development Foundation, 148
China Foundation for Poverty Alleviation, 149, 150
China Soong Ching Ling Foundation, 148–49
Churches Supporting Churches coalition, 110
civil service. *See* public administrators
civil society, 2, 146, 151; definitions of, 9, 75–76, 78; state and, 76–77, 146
climate change, 16, 40
Common Good organization, 114

communitarian social justice, 7–8, 59–61, 68–69, 96–97; impartiality doctrines of, 63–66; local capacity building and, 86–89, 170–71; partiality doctrines of, 64, 66–68, 184n20; professional ethics and, 62–64, 68–75, 83, 171, 174
communitarianism, 69–75, 79, 81; egalitarianism and, 85, 92, 96–99; liberalism and, 69–70, 74, 185n36, 185n44
community-based resilience management, 81, *82,* 97–99, 165–74; Aceh tsunami and, 124–36, 171–73; comparative analysis of, 5, 8–9, 103–5, 171–73; Fukushima Daiichi nuclear accident and, 153–63, 171–73; guidelines for, 176–77; Hurricane Katrina and, 106–22, 171, 172; Wenchuan earthquake and, 138–51, 171–73. *See also* disaster emergency management
Community Development Corporation, 119, 120
competency in community-based resilience management, 81, *82,* 97–99, 173; Aceh tsunami and, 131–36, 173; Fukushima Daiichi nuclear accident and, 160–63, 173; guidelines for, 176–77; Hurricane Katrina and, 117–22, 173; professional helpers and, 169; Wenchuan earthquake and, 145–51, 173
"complex equality," 90, 92, 93, 99
Comprehensive Vulnerability Management, 47–48; resilience doctrine of, 55–58, 57t, 81, *82*; resistance doctrine of, 48–51, 56–58, 57t; sustainability doctrine of, 51–54, 56–58, 57t. *See also* vulnerability
coping capacity, 4, 32–37, 40–42, 44–45
Cordon, Emmet, 20
Crosby, Ned, 78
Cuban hurricanes, 50
cultural competence, 118–19

cultural identity, 96–97, 165–66. *See also* arts and cultural communities
Curtin, Leah L., 95
Czech Republic floods (1997), 26

de Leon, Peter, 78
defense mechanisms, 49
Democratic Party of Japan, 153
deontology, 65
Devereux, Stephen, 34–35
Diamond, Larry Jay, 75
Dilley, Maxx, 37
disasters, 13–14; assessment of, 31–32; as community catalyst, 75–79; definitions of, 14–15; types of, 15–18, *17*
disaster emergency management (DEM), 4–9, 47–60, 58t; bottom-up approach to, 1–2, 27–30; cycle of, 23–25, *23*; history of, 18–22; process of, 22; top-down approach to, 2–3, 25–27, 30, 123. *See also* community-based resilience management
domestic violence, 142–43, 159
Durkheim, Emile, 69
Dvorkin, Ronald, 70

Earth Summit in Rio de Janeiro (1992), 52
Economic Commission for Latin America and the Caribbean, 40–42
Economic Vulnerability Index (EVI), 41
education policies: after Aceh tsunami, 128–29; after Fukushima Daiichi nuclear accident, 162; after Hurricane Katrina, 115–17, 171, 172; after Wenchuan earthquake, 138–42, 148–49, 172
egalitarianism, 85, 92, 96–99. *See also* communitarianism
Ellis, Frank, 34
empowerment, 2–3, 72, 100–101, 114, 166–67; competency and, 169; egalitarianism and, 85, 122; health care and, 121; resilience and, 42, 170, 174
environmentalism, 16, 40–41

equality, "complex," 90, 92, 93, 99
Equity and Inclusion Campaign, 114
ethics, professional, 62–64, 68–75, 83, 95, 171, 174
exclusion, 99–100. *See also* inclusion

Federal Emergency Management Agency (FEMA), 48, 120, 121, 182n5
food insecurity, 32–38
Foot, Philippa, 67
Frederickson, George, 65, 78
Freud, Sigmund, 48–49. *See also* psychotherapy
Friedrich, Carl J., 77
Fukushima Daiichi nuclear accident (2011), 1, 8–9, 104–5, 153–63, 171–73
Fukuyama, Francis, 88

Gampong, 135–36
Garfalo, Charles, 65
gender. *See* women's issues
Georgetown University, 118
Geuras, Dean, 65
globalization, 34; impact on civil service of, 122; impact on livelihoods by, 38
Godschalk, David R., 52
Good Neighbors Japan, 157
Goodin, Robert, 184n20
gotong royong (mutual assistance), 132
Gulf Opportunity Zone, 107
Guy, Mary, 65

Habitat for Humanity Japan, 161
Hansen, Gladys, 20
harm prevention, 89–94. *See also* risk reduction strategies
health care, 26, 50, 63, 95, 174; after Aceh tsunami, 134; after Fukushima Daiichi nuclear accident, 157–62; after Hurricane Katrina, 109, 121
Help Heal the Soul and Protect Traditional Culture program, 150
Honduras, 56
Horizon Education Culture Development Center (Beijing), 147

Housing Authority of New Orleans (HANO), 107
Huang Qi, 139
Huber, Evelyne, 76
Hurricane Katrina (2005), 8–9, 104, 106–22; advocacy and, 106–12, 171, 172; competency and, 117–22; inclusion and, 112–17, 172; school reform after, 115–17, 171, 172
Hyogo Framework for Action (HFA), 29–30

impartiality, 61, 63–68, 89
inclusion in community-based resilience management, 81, *82,* 99–102, 172–73; Aceh tsunami and, 127–31, 172; Fukushima Daiichi nuclear accident and, 156–60, 172; guidelines for, 175–77; Hurricane Katrina and, 112–17, 172; professional helpers and, 167–68, 172–73; Wenchuan earthquake and, 142–44, 172–73
Indo-Pakistani War (1971), 21
Indonesia: Aceh tsunami in, 8–9, 104, 124–36, 171–73; civil service of, 122–24
Indonesian Women's Association for Justice, 130
Institute of Cultural Affairs, 160–61
International Covenant on Social and Economic Rights, 107
International Monetary Fund, 41
International Panel of Climate Change, 40
International Strategy for Disaster Reduction, 15, 29
International Volunteer Center of Yamagata, 161
Internet technology, 79
Investigative News Network, 113
Iranian earthquake (2003), 28, 180n37
IsraAID, 160

Jackson, Andrew, 105
Japan: civil service of, 123–24, 151–53; Tōhoku tsunami in, 1, 8–9, 104–5, 153–63, 171, 172

Japan Association for Refugees, 159
Japan Overseas Christian Medical Cooperative Service, 161–62
Japan Volunteer Center, 163
Japanese Organization for International Cooperation in Family Planning, 159
Japanese Society for Public Administration, 152
Jewish Federation of Greater New Orleans, 120
Johnson, Chalmers, 152

Kaiser Family Foundation, 109
Kant, Immanuel, 65, 70
Kapok Community Development Research Center, 150
Kasher, Asa, 62
Kates, Robert W., 31
Kirby, Peadar, 42
Kleist, Heinrich von, 13
Kokkyo naki Kodomotachi (Children without Borders), 162
Komnas Perempuan, 130
Krueng Sabee, Indonesia, 132–33
Kymlicka, Will, 71

Landphair, Juliette, 107–8
Language Access Coalition, 114
Lapeyre, Frederic, 100
Latino organizations, 114–15, 121
Lens database, 113
Lewis, Carol, 65
Lewis, Judith A., 96
Li (Chinese merit system), 136
liberalism, 69–73; communitarianism and, 69–70, 74, 185n36, 185n44; universalism and, 70
Lisbon earthquake (1755), 13
local capacity building, 86–89
Louisiana State University, 110
Luthar, Suniya S., 55

Maas, Arthur, 77
MacArthur, Douglas, 151–52
MacIntyre, Alasdair, 69–71
Maoxian Association for Development, 150–51
Maoxian Women's Federation, 150

Mary Queen of Vietnam Church (New Orleans), 119–20
Mayday New Orleans, 108–11, 122
McClelland, David C., 98
McEntire, David A., 47
Médecins du Monde Japan, 161–62
Mexican currency crisis, 41
Mileti, Dennis S., 52–54
Ministry of International Trade and Industry (Japan), 152
Montesquieu, 69
Mozambique floods (2000), 28

National Alliance of Vietnamese American Service Agency (NAVASA), 120
National Center for Cultural Competence, 118
Natural Hazard Studies, 31
Neighborhood Partnerships Network, 112–14
Nero, Roman emperor, 18
NESRI survey, 108, 109
New Orleans. See Hurricane Katrina
New Orleans Coalition on Open Governance (NOCOG), 112, 113, 122
New Public Management, 78
Nightingale, Florence, 95
Noble, Wade, 118
nuclear power plant accidents: in Japan, 8, 104–5, 153–63, 171, 172; in Ukraine, 156
nurses, 63, 95, 121, 158–62
Nye, Joseph S., 42

Office for the Coordination of Humanitarian Affairs, 16–17
Office of Recovery Development Administration, 110
Organization for Industrial Spiritual and Cultural Advancement, 158
Oxfam Japan, 159

PADMA Indonesia, 135
Pakistan, Bhola cyclone in, 21
Pancasila (Five Principles), 123
partiality, 64, 66–68, 184n20
Peace Winds Japan (PWJ), 161

Pelayanan Advokasi untuk Keadilan dan Perdamaian (Advocacy Services for Justice and Reconciliation), 135
Pendleton Civil Service Reform Act, 105
Pflugfelder, Gregory, 1
Philippines, volcanic eruptions in, 53
Pliny the Younger, 19–20
pluralism: of distributive possibilities, 91–92; Walzer on, 98–99
Polanyi, Karl, 42–43
policy formulation, 25–30
PolicyLink, 110
Pompeii, destruction of, 18–19
Pops. Gerald, 65
Portugal, earthquake of 1755 in, 13
Potter, Nancy, 87–88
Presidential Policy Directive 8, 1
professional helpers, 1–2, 7, 165–77; advocacy by, 166–67, 171, 174–76; competency of, 169; ethics of, 62–64, 68–75, 83, 171, 174; guidelines for, 174–77; inclusion and, 167–68, 172–73, 175–77; role of, 83–86, 96–97. See also public administrators
Project on Government Oversight, 113
psychotherapy, 48–49, 157–58, 161, 162. See also health care
Pu Fei, 139
public administrators, 9, 82–83, 105–6, 165–66; of China, 123–24; ethics of, 62–64, 68–75, 83, 171, 174; guidelines for, 63–68, 173–76; of Indonesia, 122–24; of Japan, 123–24, 151–53; political engagement among, 183n5; role of, 83–86, 96–97; of United States, 105–6, 123–24. See also professional helpers
Public Affairs Research Council, 112–14
Public Insight Network, 113
Public Law Center, 113
Puentes New Orleans, 113–15
Purta Kande organization, 130
Putnam, Neil, 76

230 Index

Qiang community, 143–44, 148, 150, 172
Qin Dynasty, 136
Quarantelli, Enrico L., 14–15, 20

Radway, Laurence, 77
Rapid Evaluation and Action for Community Health in Louisiana (REACH-LA), 109, 122
Rawls, John, 70
Red Crescent organization, 21, 28, 180n37
Red Cross organization, 21, 22, 146–47, 180n37
Redding, Gordon S., 88
resilience doctrine, *17,* 55–58, 57t, 81, *82*; coping capacity and, 32–37, 40–42, 44–45; empowerment and, 42, 170, 174
resistance doctrine, 48–51, 56–58, 57t
Riely, Frank, 36
Rio de Janeiro Earth Summit (1992), 52
risk reduction strategies, *17, 23,* 24, 29–30; harm prevention by, 89–94; multidimensional categories of, 44; vulnerability and, 36–39
Roman disasters, 18–19
Rousseau, Jean-Jacques, 13, 69
Rural Economic Research Institute (RESI), 149–50
Russian currency crisis, 41

Safe Streets/Strong Communities coalition, 121
San Francisco earthquake (1906), 19–20
Sandel, Michael, 69–72
Save the Children organization, 156–57
Schmitter, Philippe C., 76
school reforms: after Aceh tsunami, 128–29; after Fukushima Daiichi nuclear accident, 162; after Hurricane Katrina, 115–17, 171, 172; after Wenchuan earthquake, 138–42, 148–49, 172

Sen, Amartya K., 33
Service for Health in Asian and African Regions, 161
Shanti Volunteer Association (SVA), 163
shari'a law, 129–30
Sichuan. *See* Wenchuan earthquake
Social Aid and Pleasure Clubs (SAPCs), 119
social justice. *See* communitarian social justice
Social Vulnerability Index (SVI), 37–38, 41, 182n50
Spatial Video Acquisition System (SVAS), 110
Stepan, Alfred, 76
Stiglitz, Joseph, 38
Stivers, Camilla, 78
Stockholm Conference on Human Environment (1972), 51
Sukarno, 123
Sullivan, Dennis T., 19
Sumatra. *See* Aceh tsunami
sustainability doctrine, 51–54, 56–58, 57t
Svara, James, 65–66
Szechwan. *See* Wenchuan earthquake

Tacitus, 18, 19
Tan Zuoren, 140
Tangshan earthquake (1976), 21–22
Taylor, Charles, 70, 71
Terra People Act Kanagawa (TPAK), 157
Tibet, 138
Titus, Roman emperor, 19
Tōhoku tsunami (2011), 1, 8–9, 104–5, 153–63, 171, 172
Tokyo Electric Power Company (TEPCO), 154, 156, 172
Tomasello, Michael, 166
tourism, after Hurricane Katrina, 117–18
trustworthiness, 87–89

Ukraine, 156
UNICEF, 130
United Nations: Department of

Economic and Social Affairs, 39; Development Fund for Women, 130–31; Economic and Social Council, 41; Economic Commission for Latin America and the Caribbean, 40–42; Environmental Programme, 40–41; International Strategy for Disaster Reduction, 15, 29; Office for the Coordination of Humanitarian Affairs, 16–17; Stockholm Conference on Human Environment of, 51
United Nations Development Programme (UNDP), 39–40
United Way of Greater New Orleans, 115
Universal Declaration of Human Rights (UDHR), 107
universalism, 70
Urban Poverty Project (UPP), 134
utilitarianism, 65

Vesuvius, Mount, 18–19
Vietnamese, of New Orleans, 119–20
Vietnamese-American Youth Leaders Association (VAYLA), 120
Vietnamese Initiatives in Economic Training (VIET), 120
vulnerability, *17,* 31–32, 38–45; assessment of, 6–7, 47–48; categories of, 44–45; coping capacity and, 32–37, 40–42, 44–45; to currency crises, 41; definitions of, 32, 39; food security as, 32–38; indices of, 37–38, 41–42, 182n50; Polyani on, 42–43; resilience doctrine of, 55–58, 57t, 81, *82;* resistance doctrine of, 48–51, 56–58, 57t; risk exposure and, 36–39; sustainability doctrine of, 51–54, 56–58, 57t. *See also* Comprehensive Vulnerability Management

Walzer, Michael, 7–8, 69–71, 79, 81–102; on cultural identity, 96–97; on egalitarianism, 85, 92, 96–99; on pluralism, 98–99; on principle of need, 122; on school reform, 117
Watts, Michael J., 35
Waugh, William L., Jr., 25
Weil, Frederick, 119
Wenchuan earthquake (2008), 8–9, 104, 138–51, 171–73
White, Gilbert F., 31
Whole Community approach, 1, 48, 182n5
Women's Empowerment Bureau (Indonesia), 131
women's issues, 168; after Aceh tsunami, 127–30; and domestic violence, 142–43, 159; after Fukushima Daiichi nuclear accident, 159–60; after Wenchuan earthquake, 142–43, 150. *See also* children's issues
World Bank, 41, 134
World Commission on Environment and Development, 52
World Conference on Natural Disaster Reduction (1994), 29
World Council on Economic Development, 51–52
World Disaster Report, 14
World Vision Japan, 158

Yangtze River flood (1931), 20
Yokohama Strategy, 29

Zeng Hongling, 140
Zuo Xiaohuan, 139

www.ingramcontent.com/pod-product-compliance
Ingram Content Group UK Ltd.
Pitfield, Milton Keynes, MK11 3LW, UK
UKHW041945140426
5217IPUK00014B/659